D1487481

METHODS OF PROTEIN SEPARATION

Volume 1

BIOLOGICAL SEPARATIONS

Series Editor: **Nicholas Catsimpoolas**
Massachusetts Institute of Technology
Cambridge, Massachusetts

Methods of Protein Separation, Volume 1
Edited by Nicholas Catsimpoolas • 1975

A Continuation Order Plan is available for this series. A continuation order will bring delivery of each new volume immediately upon publication. Volumes are billed only upon actual shipment. For further information please contact the publisher.

METHODS OF PROTEIN SEPARATION

Volume 1

Edited by
Nicholas Catsimpoolas
Massachusetts Institute of Technology
Cambridge, Massachusetts

PLENUM PRESS • NEW YORK AND LONDON

Library of Congress Cataloging in Publication Data
Main entry under title:

Methods of protein separation.

(Biological separations)
Includes bibliographies and index.
1. Proteins. 2. Separation (Technology) 3. Biological chemistry. I.
Catsimpoolas, Nicholas.

QD431.M47	574.1'9245'028	75-17684

ISBN 0-306-34601-X

©1975 Plenum Press, New York
A Division of Plenum Publishing Corporation
227 West 17th Street, New York, N.Y. 10011

United Kingdom edition published by Plenum Press, London
A Division of Plenum Publishing Company, Ltd.
Davis House (4th floor), 8 Scrubs Lane, Harlesden, London, NW10 6SE, England

CONTRIBUTORS

John R. Cann, *Department of Biophysics and Genetics, University of Colorado Medical Center, Denver, Colorado 80220*

Nicholas Catsimpoolas, *Biophysics Laboratory, Department of Nutrition and Food Science, Massachusetts Institute of Technology, Cambridge, Massachusetts 02139*

Alfred J. Crowle, *Webb-Waring Lung Institute and Department of Microbiology, University of Colorado, School of Medicine, Denver, Colorado 80220*

Pedro Cuatrecasas, *Department of Experimental Therapeutics and Medicine, The Johns Hopkins University School of Medicine, Baltimore, Maryland 21205*

James W. Drysdale, *Department of Biochemistry and Pharmacology, Tufts University School of Medicine, Boston, Massachusetts 02111*

Robert E. Feeney, *Department of Food Science and Technology, University of California, Davis, California 95616*

Eli Grushka, *Department of Chemistry, State University of New York at Buffalo, Buffalo, New York 14214*

James B. Ifft, *Department of Chemistry, University of Redlands, Redlands, California 92373*

David T. Osuga, *Department of Food Science and Technology, University of California, Davis, California 95616*

Indu Parikh, *Department of Pharmacology, The Johns Hopkins University School of Medicine, Baltimore, Maryland 21205*

Burton A. Zabin, *Bio-Rad Laboratories, Richmond, California 94804*

PREFACE

This open-end treatise on methods concerning protein separation had its beginning in an American Chemical Society symposium entitled "Contemporary Protein Separation Methods" which was held in Atlantic City, New Jersey in September 1974. The purpose of the symposium—and subsequently of the present work—was to review the available modern techniques and underlying principles for achieving one of the very important tasks of experimental biology, namely the separation and characterization of proteins present in complex biological mixtures. Physicochemical characterization was covered only as related to the parent method of fractionation and therefore involved mostly mass transport processes. Additionally, the presentation of methods for gaining insight into complex interacting protein profiles was considered of paramount importance in the interpretation of separation patterns. Finally, specific categories of proteins (e.g., chemically modified, deriving from a specific tissue, conjugated to different moieties, etc.) require meticulous trial and selection and/or modification of existing methodology to carry out the desired separation. In such cases, the gained experience provides valuable guidelines for further experimentation.

Although powerful techniques exist today for the separation and related physicochemical characterization of proteins, many biological fractionation problems require further innovations. It is hoped that the description in the present treatise of some of the available separation tools and their limitations will provide the necessary integrated background for new developments in this area.

Nicholas Catsimpoolas

Cambridge, Massachusetts

CONTENTS

Chapter 3

Immunodiffusion

Alfred J. Crowle

Chapter 4

Isoelectric Focusing in Polyacrylamide Gel

James W. Drysdale

Chapter 5
Purification of Chemically Modified Proteins
Robert E. Feeney and David T. Osuga

Chapter 6

Chromatographic Peak Shape Analysis

Eli Grushka

SEDIMENTATION AND GEL-PERMEATION CHROMATOGRAPHY OF ASSOCIATING– DISSOCIATING MACROMOLECULES:

1

THE ROLE OF LIGAND MEDIATION AND RATES OF REACTION

JOHN R. CANN

I. INTRODUCTION

Some years ago we reported that in the many applications of zone electro-phoresis to biological problems it is imperative that cognizance be taken of the fact that multiple zones need not necessarily indicate inherent hetero-geneity. Thus, both theoretical calculations (Cann and Goad, 1965) and

JOHN R. CANN, Department of Biophysics and Genetics, University of Colorado Medical Center, Denver, Colorado, 80220. This work was supported in part by Research Grant 5R01 HL 13909-22 from the National Heart and Lung Institute, National Institutes of Health, United States Public Health Service. This publication is No. 588 from the Department of Biophysics and Genetics, University of Colorado Medical Center, Denver, Colorado 80220.

experimentation (Cann, 1966) revealed that, under appropriate conditions, a single macromolecule interacting reversibly with an uncharged constituent of the solvent medium can give two zones despite instantaneous establishment of equilibrium. Moreover, a single macromolecule, which isomerizes reversibly at rates comparable to the rate of electrophoretic separation of the isomers, can give three zones. Accordingly, it was emphasized that unequivocal proof of inherent heterogeneity is afforded only by isolation of the various components. Since that time, we have extended these concepts to include the sedimentation and gel-permeation chromatography of ligand-mediated associating–dissociating macromolecules and some representative kinetically controlled interactions. A review of these advances and their implications for the separation of proteins and the characterization of biologically important interactions is given below.

II. THEORETICAL FORMULATION

While a detailed formulation of the theory of transport of interacting systems along with its several approximations and ramifications seems inappropriate, it is instructive to point out the salient features of the calculations. Consider, for example, the reversible reaction (reaction I)

$$mM + nX \rightleftharpoons M_mX_n \tag{I}$$

in which a macromolecule M associates into an m-mer, with the mediation of a small ligand molecule or ion X of which a fixed number n are bound into the complex. The set of conservation equations for sedimentation in a sector-shaped ultracentrifuge cell takes the form

$$\frac{\partial(C_1 + mC_2)}{\partial t} = \frac{1}{r}\frac{\partial}{\partial r}\left[\left(D_1\frac{\partial C_1}{\partial r} - C_1s_1\omega^2r\right)r + m\left(D_2\frac{\partial C_2}{\partial r} - C_2s_2\omega^2r\right)r\right]$$

$$\tag{1}$$

$$\frac{\partial(nC_2 + C_3)}{\partial t} = \frac{1}{r}\frac{\partial}{\partial r}\left[n\left(D_2\frac{\partial C_2}{\partial r} - C_2s_2\omega^2r\right)r + \left(D_3\frac{\partial C_3}{\partial r} - C_3s_3\omega^2r\right)r\right]$$

in which C designates the molar concentration; D, the diffusion coefficient; s, the sedimentation coefficient; ω, the angular velocity; r, the radial distance; and t, the time. The subscripts 1, 2, and 3 designate M, M_nX_n, and X, respectively. These equations conserve total macromolecule and total small molecule by taking into account diffusion and transport in the centrifugal field. A third equation must express the effect of reaction I on the concentrations. For the limiting case of rates of reaction so fast that, in effect, there is local equilibrium among the interaction species, this equation is

$$C_2 = KC_1^m C_3^n \qquad (2)$$

where K is the equilibrium constant of the reaction. Solution of the simultaneous equations (1) and (2) gives the sedimentation pattern of the interacting system; and a rapid and accurate calculation has been developed for their numerical solution on a high-speed electronic computer (Goad, 1970). Although most of the calculations have been for constant s and D, others have made allowance for the concentration dependence of the transport coefficients (Cann and Goad, 1972). The logic of the calculation is as follows. Suppose the centrifuge cell is divided into a number of discrete segments. Given the initial equilibrium concentrations of the several species in each segment, we can calculate the change in the distribution of material during a short interval of time Δt due to transport. It is assumed that there is no reequilibration by interaction during Δt, each species migrating independently from its distribution at the beginning of the interval. After the concentrations have been advanced, equilibrium is recalculated. That is, for known constituent concentrations of macromolecule and ligand computed from the concentrations of the several species as changed by transport, new equilibrium concentrations are calculated by applying the law of mass action [equation (2)]. We then compute the change in this new distribution of material due to transport over the next Δt, recalculate the equilibrium, and so on, constructing the entire evolution of the distribution of material along the centrifuge cell from the initial conditions. Either analytical or zone sedimentation patterns can be calculated depending upon whether the initial and boundary conditions correspond to the moving-boundary or the zonal mode of mass transport. Also, the computer code is so written as to be applicable to the ligand-mediated dissociation reaction

$$M_m + mnX \rightleftharpoons mMX_n \qquad (II)$$

An important simplification of the foregoing calculation is founded on transport equations which make the rectilinear and constant-field approximations. These approximations assume rectilinear rather than radial motion of the sedimenting molecules and constant rates of sedimentation. Actually, the constant-field approximation becomes exact for zone sedimentation through a preformed linear density gradient in the preparative ultracentrifuge. This is so because macromolecules sediment with constant velocities under the conditions described by Martin and Ames (1961).

The rectilinear and constant-field approximations have been applied to a number of interactions including the kinetically controlled, ligand-mediated dimerization reaction (reaction set III)

$$M + X \underset{k_2}{\overset{k_1}{\rightleftharpoons}} MX \qquad K_1 = k_1/k_2 \qquad\qquad \text{(IIIa)}$$

$$M + MX \underset{k_4}{\overset{k_3}{\rightleftharpoons}} M_2X \qquad K_2 = k_3/k_4 \qquad\qquad \text{(IIIb)}$$

where the specific rate constants k_1 and k_3 have the dimensions $\text{M}^{-1} \text{sec}^{-1}$ and k_2 and k_4 have the dimension sec^{-1} (Cann and Kegeles, 1974). The conservation equations for independent transport of the several species take the form

$$\frac{\partial C_i}{\partial t} = D_i \frac{\partial^2 C_i}{\partial x^2} - v_i \frac{\partial C_i}{\partial x} \qquad\qquad (3)$$

where v_i is the constant driven velocity. Also $v_i = s_i \omega^2 \bar{x}$, in which $\bar{x}$ is an average position in the centrifuge cell. In order to minimize truncation error these equations are solved in a frame of reference moving with the average velocity of macromonomer and dimer, $\bar{v} = (v_1 + v_2)/2$, which introduces the new position variable, $x' = x - \bar{v}t$. Equations (3) are transformed into the moving coordinate system simply by making the replacements $x' \rightarrow x$ and $v_2 - \bar{v} \rightarrow v_i$. The algorithm for reaction set III is as follows: (1) calculate the change in the initial equilibrium distribution of material during Δt due to independent transport of each species. (2) Calculate new equilibrium concentrations by imposing the law of mass action. (3) Relax the concentrations for a length of time Δt from their values as changed by transport toward their new equilibrium values using the theory of relaxation kinetics. (4) Use the relaxed concentrations as the starting distribution of material for the next time cycle of transport which is followed, in turn, by relaxation of ligand binding, relaxation of the dimerization reaction, etc. This recursive calculation constructs the evolution of the sedimentation pattern.

Less complex kinetically controlled interactions are more conveniently treated by incorporating the chemical kinetics directly into the conservation equation for each species (Cann and Oates, 1973). An example is the irreversible dissociation of a macromolecule P into m hydrodynamically identical subunits M

$$P \xrightarrow{k} mM \qquad\qquad\qquad\qquad \text{(IV)}$$

The system of equations whose solution gives the sedimentation pattern is

$$\frac{\partial C_1}{\partial t} = D_1 \frac{\partial^2 C_1}{\partial x^2} - v_1 \frac{\partial C_1}{\partial x} - kC_1$$

$$\frac{\partial C_2}{\partial t} = D_2 \frac{\partial^2 C_2}{\partial x^2} - v_2 \frac{\partial C_2}{\partial x} + mkC_1 \tag{4}$$

where the subscripts 1 and 2 designate P and M, respectively. These equations conserve mass by taking into account diffusion, sedimentation, and the chemical reaction which occurs during transport. One proceeds by assuming that the diffusion coefficients vary inversely with the cube root of the molecular weight, $D_2 = m^{1/3} D_1$, and by introducing the dimensionless time and position variables

$$t_0 = kt \tag{5}$$

$$x_0 = (k/D_1)^{1/2} x - (k/D_1)^{1/2} \bar{v} t \tag{6}$$

which transforms equations (4) into

$$\frac{\partial C_1}{\partial t_0} = \frac{\partial^2 C_1}{\partial x_0{}^2} + \frac{1}{2\alpha^{1/2}} \frac{\partial C_1}{\partial x_0} - C_1$$

$$\frac{\partial C_2}{\partial t_0} = m^{1/3} \frac{\partial^2 C_2}{\partial x_0{}^2} - \frac{1}{2\alpha^{1/2}} \frac{\partial C_2}{\partial x_0} + mC_1 \tag{7}$$

where

$$\alpha = \frac{D_1 k}{(v_2 - v_1)^2} \tag{8}$$

Numerical solution of the simultaneous equations (7) for the constituent concentration, $4C_1 + C_2$, as a function of x_0 for increasing t_0 over a range of values of the parameter α provides a dimensionless representation of the effect of time of sedimentation, rate of reaction, and difference in sedimentation velocity between the macromolecule and its subunits upon the shape of the sedimentation pattern. Computations have been made for the case in which the macromolecule exists entirely in state P at the start of the sedimentation experiment. Essentially, the same mathematical formulation has been applied to irreversible dimerization

$$2M \xrightarrow{k} M_2 \tag{V}$$

and both irreversible and reversible dissociation of a complex

$$C \underset{k_2'}{\overset{k_1'}{\rightleftharpoons}} A + B \tag{VI}$$

where $k_2'/k_1' \geq 0$.

Finally, we have examined (Cann, 1973) the sedimentation behavior of a number of noncooperative ligand-mediated interactions for which equilibration is rapid. These include the set of sequential and simultaneous interactions

$$
\begin{aligned}
M + X &\rightleftharpoons MX & K_1 \\
MX + M &\rightleftharpoons M_2X & K_2 \\
2MX &\rightleftharpoons M_2X_2 & K_3
\end{aligned}
\tag{VII}
$$

the sequential monomer–tetramer reaction

$$
\begin{aligned}
M + X &\rightleftharpoons MX & K_1 \\
4MX &\rightleftharpoons M_4X_4 & K_2
\end{aligned}
\tag{VIII}
$$

the sequential and progressive tetramerization

$$
\begin{aligned}
M + X &\rightleftharpoons MX & K_1 \\
2MX &\rightleftharpoons M_2X_2 & K_2 \\
M_2X_2 + MX &\rightleftharpoons M_3X_3 & K_3 \\
M_3X_3 + MX &\rightleftharpoons M_4X_4 & K_4
\end{aligned}
\tag{IX}
$$

$$
K_2 = K_3 = K_4 \equiv K
$$

and the dissociation of a dimer driven by binding of ligand to the monomer

$$
\begin{aligned}
M_2 &\rightleftharpoons 2M & K_1 \\
M + X &\rightleftharpoons MX & K_2
\end{aligned}
\tag{X}
$$

In each case, a recursive calculation was used in which independent transport to the rectilinear and constant-field approximations is followed by reequilibration, etc. The conservation equations take the form of equations (3) and are solved in a moving coordinate system x'. The same logic was applied to small-zone gel-permeation chromatography (Ackers, 1969) in the total-column frame of reference (Zimmerman and Ackers, 1971). For this calculation the diffusion coefficients in the conservation equations were replaced by axial dispersions, and the velocity of a given species was set equal to the flow rate divided by its partition cross section (Zimmerman et al., 1971).

As for notation over and above that already given, molar concentrations bear the following subscripts: For reaction set III the subscripts 1, 2, and 3 designate total monomer, both uncomplexed and complexed with ligand, dimer, and unbound ligand, respectively. In reaction V, 1 and 2 designate M and P. In reaction VI, 1, 2, and 3 designate C, A, and B. In reaction set VIII, 1, 4, and 5 designate total monomer, tetramer, and unbound ligand. In reaction set IX, 1, 2, 3, 4, and 5 designate total monomer, dimer, trimer, tetramer, and unbound ligand. Initial concentrations bear a tenfold-larger subscript; e.g., in reaction set III, C_{10} and C_{20} designate initial equilibrium concentrations of total monomer and of dimer, respectively, so that the initial constituent concentration of macromolecule is $C_{10} + 2C_{20}$.

For purposes of discussion it is convenient to characterize reaction set III in terms of the ratio R_{10} of the initial equilibrium concentration of MX to that of M so that

$$K_1 = R_{10}/C_{30} \tag{9a}$$

$$K_2 = (C_{20}/C_{10}{}^2) \cdot [(1 + R_{10})^2/R_{10}{}^2] \tag{9b}$$

similarly, for reaction set VIII

$$K_1 = R_{10}/C_{50} \tag{10a}$$

$$K_2 = (C_{40}/C_{10}{}^4) \cdot [(1 + R_{10})^4/R_{10}{}^4] \tag{10b}$$

Reaction set IX is characterized by

$$K_1 = R_{10}/C_{50} \tag{11}$$

and the assigned value of K.

The computed analytical sedimentation patterns are displayed as plots of concentration gradient against position; e.g., $\delta(C_1 + 2C_2)/\delta x'$ vs. x' for reaction set III. Zone patterns are plots of constituent concentration vs. position; e.g., in the case of reaction set VIII, $C_1 + 4C_4$ vs. x' for sedimentation and $C'_T \equiv C'_1 + 4C'_4$ vs. x' for gel-permeation chromatography, where the primed concentrations indicate the total-column frame of reference (Zimmerman and Ackers, 1971). The vertical arrows in most of the patterns indicate where the peaks would have been located had sedimentation or chromatography been carried out on a mixture of noninteracting macromolecules having the same transport parameters as the reactant and product of the reacting system.

III. RESULTS

As is evident from the foregoing, the sedimentation behavior of several different types of macromolecular interactions have been examined. Broadly speaking, these can be grouped into two classes: (1) ligand-mediated association–dissociation reactions with either instantaneous or kinetically controlled reequilibration during differential transport of the various molecular species, and (2) kinetically controlled, nonmediated interactions such as the irreversible dissociation of a protein into its subunits and the dissociation of an enzyme complex either irreversibly or reversibly. A summary of the more important results of the calculations for the two classes of interactions follows.

A. Ligand-Mediated Association–Dissociation Reactions

Let us first consider the type of ligand-mediated macromolecular association schematized by reaction I. While analytical sedimentation patterns have been calculated (Cann and Goad, 1972; Cann, 1970) for both dimerization and tetramerization, the results for dimerization (reaction I with $m = 2$ and $n = 1$ or 6) are, in certain respects, the more revealing with regard to fundamental principles. A particularly significant result is that ligand-mediated dimerization can give rise to a well-resolved bimodal reaction boundary (Figure 1B) despite the instantaneous reequilibrium. Resolution of the two peaks depends upon the production and maintenance of concentration gradients of unbound ligand along the centrifuge cell by reequilibration during differential transport of macromonomer and dimer; the higher the cooperativeness of the interaction (i.e., the larger n), the better the resolution. The peaks correspond to different equilibrium mixtures and not simply to monomer and dimer. Resolution of the peaks is in contradistinction to the behavior predicted by the Gilbert theory (Gilbert, 1955, 1959) and observed experimentally (Field and O'Brien, 1955; Field and Ogston, 1955) for nonmediated dimerization

$$2M \rightleftharpoons M_2 \qquad\qquad (XI)$$

Systems of this sort always give sedimentation patterns showing a single peak with a weight-average sedimentation velocity when reequilibration is

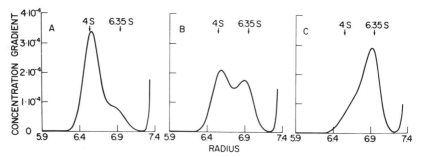

FIGURE 1. Theoretical analytical sedimentation patterns for the ligand-mediated dimerization reaction $2M + 6X \rightleftharpoons M_2X_6$. Dependence of boundary shape upon percent dimerization, $K = 4.57 \times 10^{29}$ M^{-7}: (A) 25% dimerization, $C_{30} = 3.89 \times 10^{-5}$ M; (B) 50%, 5×10^{-5} M; (C) 75%, 6.74×10^{-5} M. The following values of macromolecular concentrations and parameters were chosen to approximate a protein of molecular weight 60,000: $C_{10} + 2C_{20} = 14 \times 10^{-5}$ M, $s_1 = 4$ S, $s_2 = 6.35$ S, $D_1 = 6 \times 10^{-7}$ cm^2 sec^{-1}, $D_2 = 4.76 \times 10^{-7}$ cm^2 sec^{-1}. For the small ligand molecule: $s_3 = 0.1$ S, $D_3 = 10^{-5}$ cm^2 sec^{-1}. Rotor speed, 60,000 rev/min. Time of sedimentation 6431 sec. From Cann (1970).

rapid. Of course, at sufficiently high ligand concentration ligand-mediated dimerization will also give patterns that show a single peak. In the limit where for any reason the concentration of unbound ligand along the centrifuge cell is not significantly perturbed by the reaction during differential sedimentation of the macromolecular species, the system effectively approaches the case of nonmediated dimerization considered by Gilbert.

Although the described sedimentation behavior was originally predicted based on calculations which assume constant transport parameters (Cann, 1970), essentially the same conclusions derive from calculations which take cognizance of the hydrodynamic dependence of the sedimentation coefficients upon concentration (Cann and Goad, 1972).

The theoretical prediction of bimodal reaction boundaries found experimental verification in studies on the dimerization of tubulin through the mediation of one vinblastine molecule (Weisenberg and Timasheff, 1970) and on the reversible dimerization (hexamer–dodecamer reaction) of New England lobster hemocyanin mediated by the binding of 4–6 Ca^{2+} and 2–4 H^+ which occurs when the pH is lowered from above 9.6 to below 9.2 (Morimoto and Kegeles, 1971; Tai and Kegeles, 1971). Subsequently, however, stopped-flow light-scattering measurements revealed (Kegeles and Tai, 1973) that the hexamer–dodecamer reaction of lobster hemocyanin does not equilibrate instantaneously, the half-times of reaction being of the order of 40–100 sec depending upon conditions. This finding prompted a theoretical investigation into the effect of chemical kinetics upon the shape of the reaction boundary corresponding to 50% dimerization of a protein with the molecular weight of lobster hemocyanin (Cann and Kegeles, 1974). Comparative calculations were made for the kinetically controlled, ligand-mediated dimerization reaction set III and for the same set of reactions subject to instantaneous reequilibration. These, in turn, were compared with similar calculations for nonmediated dimerization reaction XI.

Representative sedimentation patterns for reaction set III are displayed in Figure 2. The shape of the reaction boundary for instantaneous reequilibration (Figure 2A) depends upon the initial concentration of unbound ligand C_{30}. For sufficiently low C_{30} [i.e., sufficiently strong ligand binding; see equation (9a)] the pattern is well resolved into two peaks, but the gradient never goes to the baseline between the peaks. Increasing C_{30} while holding R_{10} and the percent dimerization constant [see equations (9a) and (9b)] causes progressive loss of resolution, until at the highest value of C_{30} the pattern shows a single peak whose shape approaches that found for nonmediated dimerization when reequilibration is instantaneous. We now inquire how kinetic control affects the shape of the boundary at a given value of C_{30}. Toward this end we make the reasonable assumption that ligand-binding *per se* (reaction IIIa) equilibrates rapidly so that the monomer–dimer

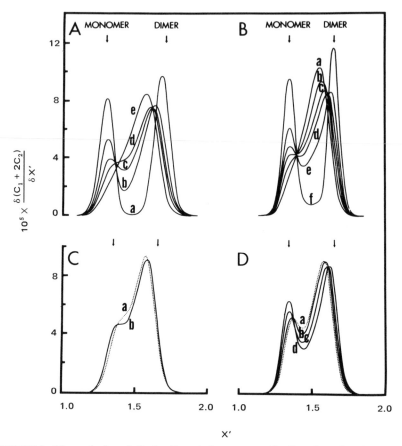

FIGURE 2. Theoretical analytical sedimentation patterns for ligand-mediated dimerization (Reaction set III). The percentage of dimerization is 52.7%. A—Instantaneous reequilibration during differential transport of the several species: (a) $C_{30} = 10^{-7}$ M; (b) 5×10^{-7} M; (c) 7.5×10^{-7} M; (d) 10^{-6} M; and (e) 2×10^{-6} M. B—Kinetically controlled interaction for $C_{30} = 2 \times 10^{-6}$ M: (a) $k_4 = 3.6 \times 10^{-2}$ sec^{-1} (half-time, 19 sec); (b) 6×10^{-3} sec^{-1}; (c) 3×10^{-3} sec^{-1}; (d) 2×10^{-3} sec^{-1}; (e) 10^{-3} sec^{-1}; and (f) 10^{-4} sec^{-1} (half-time, 6.9×10^3 sec). C—Comparison of pattern for instantaneous reequilibration [curve (a)] with the pattern for kinetically controlled interaction with $k_4 = 1.2 \times 10^{-2}$ sec^{-1}, i.e., half-time of 58 sec [curve (b)]; $C_{30} = 10^{-6}$ M. D—Comparison of rapidly equilibrating and kinetically controlled interactions for $C_{30} = 7.5 \times 10^{-7}$ M: (a) instantaneous reequilibration; (b) $k_4 = 1.2 \times 10^{-2}$ sec; (c) 3×10^{-3} sec^{-1}; (d) 1.5×10^{-3} sec^{-1}. $C_{10} + 2C_{20} = 2.326 \times 10^{-5}$ M; $R_{10} = 9.743 \times 10^{-3}$; see equations (9a) for K_1; $K_2 = 5.625 \times 10^6$ M^{-1}. For kinetically controlled interactions: $k_2 = 10^4$ sec^{-1}, with k_1 depending upon C_{30} (see equation (9a) and recall that $k_1 = k_2 K_1$); $k_3 = k_4 K_2$. Also $s_1 = 17$ S, $s_2 = 25$ S, $s_3 = 0$, $\bar{x} = 6.6$ cm, 60,000 rev/min, time of sedimentation $= 2 \times 10^3$ sec in A and 1.5×10^3 sec in B and C. $D_3 = 10^{-5}$ cm^2 sec^{-1}; mathematically effective diffusion coefficients of macromolecule: monomer, 6.21×10^{-7} cm^2 sec^{-1}; dimer, 5.31×10^{-7} cm^2 sec^{-1}. From Cann and Kegeles (1974).

interconversion (reaction IIIb) becomes rate limiting, the definitive kinetic parameter being the half-time of dissociation of the dimer.

The effect of chemical kinetics upon the shape of the sedimentation pattern for $C_{30} = 2 \times 10^{-6}$ M is illustrated by the family of patterns displayed in Figure 2B. At this concentration of ligand the pattern is unimodal for instantaneous reequilibration (curve e in Figure 2A). The pattern for a half-time of dissociation equal to 19 sec (curve a in Figure 2B) is almost indistinguishable from the pattern for instantaneous reequilibration; and even though a half-time of about 60 sec distorts the shape of the peak somewhat, it is possible that in practice such a pattern might be misinterpreted as indicative of very rapid equilibration. Upon further increase in the half-time of reaction, resolution into two peaks ensues and becomes increasingly sharp as the limit is approached wherein negligible interconversion occurs during the course of sedimentation. Since the calculations are for a ligand concentration at which the pattern for instantaneous reequilibration approaches that for nonmediated dimerization, it is understandable that the described effect of chemical kinetics is strikingly similar to the results of a comparable set of calculations for nonmediated dimerization. As for the effect of chemical kinetics at lower ligand concentrations where bimodal reaction boundaries obtain for instantaneous reequilibration, decreasing the rates of reaction merely sharpens the resolution of the peaks (Figure 2D). Although these calculations are for a macromonomer with the molecular weight of lobster hemocyanin, the results also apply qualitatively to smaller proteins whose sedimentation behavior will be somewhat less sensitive to kinetic control owing to slower sedimentation velocities.

The foregoing results admit the important generalization that conclusions concerning the sedimentation behavior of ligand-mediated interactions in the limit of instantaneous establishment of equilibrium evidently are valid for kinetically controlled interactions characterized by half-times of reaction as long as 20–60 sec, and the same can be said for nonmediated dimerization. Also of considerable practical importance is the finding that for half-times of dissociation of dimer less than about 200 sec, resolution of the reaction boundary into two peaks can occur only if dimerization is ligand-mediated.

While our discussion has so far dealt only with analytical sedimentation, zone sedimentation has received even greater attention (Cann and Goad, 1970; Cann, 1973) since it is so widely employed in biological research, particularly when identification and separation of macromolecules is the overriding consideration. Thus, it is of considerable practical significance that the results for zone sedimentation are in certain respects analogous to those for analytical sedimentation. Consider, for example, the band sedimentation patterns computed for dimerization mediated by the binding of six ligand molecules into the complex (reaction I with $m = 2$ and $n = 6$). The results displayed in Figure 3 show that, for appropriate choice of ligand concentration, the computed band pattern displays a well-resolved bimodal reaction zone. Increasing the initial ligand concentration, holding the percent dimerization constant, results in progressive coalescence of the two peaks

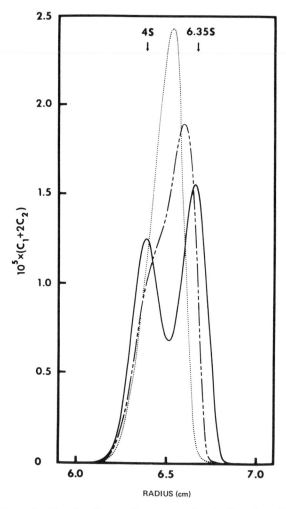

FIGURE 3. Theoretical band sedimentation pattern for the ligand-mediated dimerization reaction, $2M + 6X \rightleftharpoons M_2X_6$. 50% dimerization, (————) $C_{30} = 10^{-6}$ M, (— — —) 10^{-5} M, (· · · · ·) 5×10^{-5} M. Time of sedimentation = 4740 sec at 60,000 rev/min, $s_3 = 0.15$ S, other parameters are as in Figure 1. Unbound ligand was initially distributed uniformly along the centrifuge cell; a bimodal zone of virtually the same shape as shown here for $C_{30} = 10^{-6}$ M was also obtained when the computation was for unbound ligand initially present only in the starting zone. From Cann and Goad (1970). Copyright 1970 by the American Association for the Advancement of Science.

with concomitant drift in the sedimentation velocities toward a value close to the weight average until resolution disappears entirely at the highest concentration. The mean sedimentation velocity of the unimodal zone shown at the highest ligand concentration decreases slightly during the course of sedimentation owing to some dissociation of the dimer within the spreading zone.

The latter behavior assumes major proportions in the case of ligand-mediated tetramerization (reaction I with $m = 4$ and $n = 4$). In fact, the strong dependence of tetramerization upon total protein concentration prevents resolution of the spreading zone into two peaks. Moreover, the mean sedimentation velocity of the predicted unimodal zone decreases rapidly and continuously from the weight-average value to a value approaching that of the monomer as the zone migrates down the centrifuge cell (Figure 4A). This finding has important implications for experiments on allosteric enzymes. Thus, if circumstances attending an unusual behavior of a unimodal zone sedimenting through a preformed density gradient in the preparative ultracentrifuge (e.g., increasing sedimentation coefficient with increasing protein concentration in the known presence of ligand) indicates ligand-mediated association–dissociation, the system should be examined by band sedimentation in the analytical instrument. In the case of band sedimentation the instantaneous sedimentation coefficient can be extrapolated to zero time thereby characterizing the interaction quantitatively. In this light, the facilitated association of carbamyl phosphate synthetase by positive allosteric effectors such as inosine monophosphate or ornithine (Anderson and Marvin, 1970) bears reinvestigation. Another important finding relates to the distribution of ligand along the centrifuge cell. Whereas the zone of macromolecule is unimodal, the distribution of total ligand is bimodal (Figure 4B) provided that initially the free ligand is distributed uniformly throughout the centrifuge cell. The bimodal distribution of total ligand can be understood as follows: As the tetramer dissociates it releases ligand which remains behind the advancing zone. The sum of the broad peak of released ligand, the peak of ligand bound to the remaining tetramer, and the distorted background of unbound ligand is bimodal. Careful note should be taken of the difference between the distribution of macromolecule and total ligand, since it is not uncommon in practice to follow zone sedimentation in the preparative ultracentrifuge by analysis of fractions for specific ligand (Gilbert and Müller-Hill, 1966). This is often the only method available for detecting a specific protein in partially purified cellular extracts; but a bimodal distribution of ligand need not necessarily indicate inherent heterogeneity of the specific protein.

A different behavior is shown by ligand-mediated dissociation of a tetramer (reaction II with $m = 4$ and $n = 4$) which can give bimodal reaction zones. This contrasting behavior relates to the fact that total ligand concen-

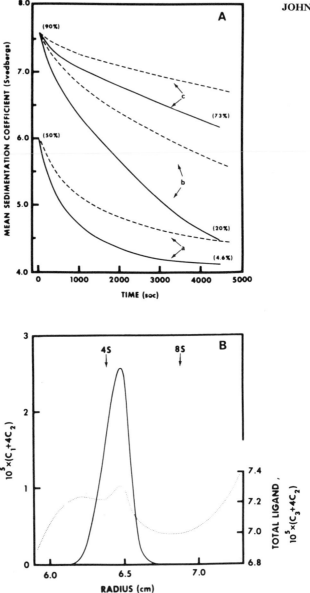

FIGURE 4. Theoretical band sedimentation behavior for cooperative ligand-mediated tetramerization, $4M + 4X \rightleftharpoons M_4X_4$. (A)—Dependence of mean sedimentation coefficient upon time of sedimentation: (———) instantaneous sedimentation coefficients computed from first moments at about 470 sec intervals; (– – – –) sedimentation coefficients computed from the first moment of the band at a given time and the first moment of the initial band. Curves (a) were computed for 50% tetramer and $C_{30} = 7 \times 10^{-5}$ M throughout the cell; curves (b) and (c) for 90% tetramer and $C_{30} = 1.4 \times 10^{-5}$ M in the initial band only or throughout the cell, respectively. Percent tetramer in initial and

tration as well as total macromolecule concentration diminish as the zone spreads. The decrease in total ligand concentration produces the asymmetry between mediated association and mediated dissociation; in the one case ligand promotes association, while in the other, ligand promotes dissociation. Another distinctive feature of the dissociation reaction becomes evident in the limit of high ligand concentration. In this limit the reaction zone becomes unimodal, and its sedimentation velocity decreases with time owing to dissociation of the tetramer as the total macromolecule concentration decreases in the spreading zone. Under these conditions the total ligand pattern also shows a single peak which matches the zone of macromolecule both in shape and velocity.

The above considerations are for zone sedimentation of reacting systems which are cooperative with respect to ligand (reactions I or II with $n > 1$); and the question remains whether noncooperative interactions can give bimodal reaction zones. Accordingly, theoretical zone sedimentation patterns have been computed (Cann, 1973) for a variety of noncooperative, ligand-mediated, association–dissociation reactions (reaction sets VII, VIII, IX, and X) for which reequilibrium is instantaneous. Under appropriate conditions each of these interactions can show bimodal reaction zones, even the tetramerization reaction sets VIII and IX (Figure 5), which is in contradistinction to the case of tetramerization mediated by cooperative interaction with ligand. This dichotomy resides in the relative insensitivity of the noncooperative interaction to the decreasing concentration of total ligand within the spreading zone as it migrates down the centrifuge tube. With the exception of tetramerization and high-order association reactions mediated by cooperative interaction with ligand (reaction I with $m \geq 4$ and $n \geq 4$), the generalization seems justified that rapidly equilibrating ligand-mediated interactions, in general, have the potential for showing bimodal zones irrespective of reaction mechanism subject to the provisos that overall ligand binding is sufficiently strong to generate large gradients of unbound ligand along the sedimentation column but that binding to the reactant itself is not so strong that in effect the system behaves like the analogous nonmediated interaction. A dramatic illustration of the latter proviso is provided by comparison of sedimentation patterns a and e in Figure 5A for reaction set VIII. For $R_{10} = 0.10$, ligand binding to the monomer is relatively weak, and the zone

---

final band of macromolecule shown in parentheses. (B)—Band sedimentation pattern after 4250 sec: (———) macromolecule concentration; ($\cdots$) total ligand concentration; initially 50% tetramer; $C_{30} = 7 \times 10^{-5}$ M; unbound ligand initially distributed uniformly along the centrifuge cell. $C_{10} + 4C_{20} = 14 \times 10^{-5}$ M; $s_1 = 4$ S; $s_2 = 8$ S; $s_3 = 0.15$ S; $D_1 = 6 \times 10^{-7}$ cm^2 sec^{-1}; $D_2 = 3.8 \times 10^{-7}$ cm^2 sec^{-1}; $D_3 = 10^{-5}$ cm^2 sec^{-1}; 60,000 rev/min. From Cann and Goad (1970). Copyright 1970 by the American Association for the Advancement of Science.

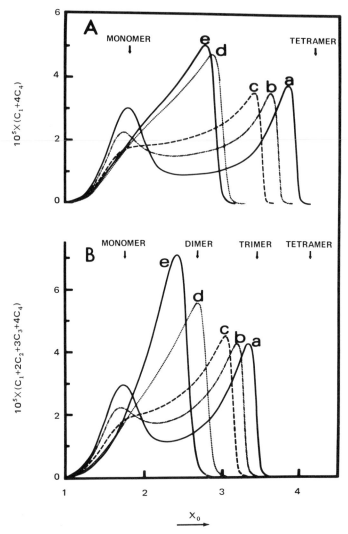

FIGURE 5. Factors governing the shape of the theoretical zone sedimentation pattern for noncooperative, ligand-mediated tetramerization. A—Reaction set VIII, 50% tetramerization with $R_{10} = 0.10$ for patterns (a), (b), (c), and (d) and $R_{10} = 10.00$ for pattern (e). Pattern (a), $C_{50} = 5 \times 10^{-7}$ M; (b) 2×10^{-6} M; (c) 5×10^{-6} M; (d) 7×10^{-5} M; and (e) 5×10^{-7} M [see equations (10a) and (10b) for K_1 and K_2]. B—Reaction set IX, 50% association with $R_{10} = 0.10$ and $K = 1.72 \times 10^5$ M^{-1} for patterns (a), (b), (c), and (d) and $R_{10} = 10.00$ and $K = 5.10 \times 10^3$ M^{-1} for pattern (e): pattern (a) $C_{50} = 5 \times 10^{-7}$ M; (b) 2×10^{-6} M; (c) 5×10^{-6} M; (d) 5×10^{-5} M; and (e) 5×10^{-5} M [see equation (11) for K_1]. Initial constituent concentration of macromolecule equals 14×10^{-5} M, $s_1 = 4$ S, $s_2 = 6.35$ S, $s_3 = 8.32$ S, $s_4 = 10.08$ S, $s_5 = 0.1$ S, $D_1 = 6 \times 10^{-7}$ cm^2 sec^{-1}, $D_2 = 4.76 \times 10^{-7}$ cm^2 sec^{-1}, $D_3 = 4.16 \times 10^{-7}$ cm^2 sec^{-1}, $D_4 = 3.78 \times 10^{-7}$ cm^2 sec^{-1}, $D_5 = 10^{-5}$ cm^2 sec^{-1}. 60,000 rev/min; $\bar{x} = 6.6$ cm; time of sedimentation $= 1.5 \times 10^4$ sec. From Cann (1973).

(pattern a) is bimodal. When the strength of binding is increased 100-fold by increasing the value of R_{10} from 0.10 to 10.0, holding C_{50} and percent tetramerization constant [see equations (10a) and (10b)], the system collapses to the nonmediated tetramerization reaction

$$4M \rightleftharpoons M_4 \qquad \text{(XII)}$$

and thus gives a unimodal zone (pattern e).

We have long maintained that the results described above for zone sedimentation and elsewhere (Cann and Goad, 1968) for zone electrophoresis can be applied to chromatography with only quantitative reservations. To bring this assertion into focus, our calculations on reaction set VIII have been extended in a quantitative fashion to include gel-permeation chromatography on Sephadex. As illustrated in Figure 6A, the predicted chromatographic behavior is qualitatively similar to the sedimentation behavior depicted in Figure 5A. It is particularly interesting that increasing the flow rate (Figure 6B) causes progressive coalescence of the two peaks in the broadening reaction zone. This is due to progressive increase in the axial dispersions of the several species, especially the increase in the axial dispersion of the ligand, which tends to smooth its concentration gradients through the zone. Although the chromatographic patterns were computed for the total-column frame of reference and are thus directly related to the column-scanning mode of data acquisition (Ackers, 1969), the elution profile will show the same features. Accordingly, the same precaution that was emphasized previously for electrophoresis and sedimentation (Cann and Goad, 1965, 1970; Cann 1970; Cann and Oates, 1973) also applies to gel filtration; namely, fractions must be rechromatographed to see if they run true. Otherwise, bimodality due to interactions could be misinterpreted in terms of inherent heterogeneity.

B. Kinetically Controlled Nonmediated Interactions

Our interest in kinetically controlled nonmediated interactions was aroused by discussions with colleagues concerning the electrophoresis, sedimentation, and gel filtration of enzyme systems and other reacting macromolecules. In its simplest terms, the most frequently posed question was "Will the mass-transport pattern of a macromolecule undergoing slow and irreversible reaction during the time course of the experiment show two peaks or a single, skewed peak with a long trailing or leading edge?" In order to answer this question, both moving-boundary and zone patterns were calculated for irreversible isomerization, irreversible dissociation of a macromolecule into its hydrodynamically identical subunits (reaction IV), irreversible dimerization (reaction V), and irreversible and reversible dissociation of

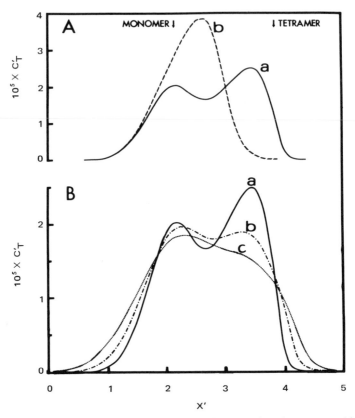

FIGURE 6. Factors governing the shape of the gel-permeation chromatographic profile computed for Reaction set VIII on Sephadex G-200R, 50% tetramerization. A—Dependence of shape on R_{10} for flow rate $F = 1.2$ ml/hr: (a) $R_{10} = 0.10$; (b) $R_{10} = 10.00$. B—Dependence of shape on F for $R_{10} = 0.10$: (a) $F = 1.2$ ml/hr; (b) $F = 5$ ml/hr; (c) $F = 9.6$ ml/hr. Ligand was initially present only in the starting zone; where tested (conditions in A) virtually the same profiles were obtained when unbound ligand was initially distributed throughout the chromatographic column. The times are such that the volume passed ($V = Ft$) is 5 ml for all curves. $C'_{10} + 4C'_{40} = 14 \times 10^{-5}$ M; $C'_{50} = 5 \times 10^{-7}$ M. Monomer, 17,000 daltons. From Cann (1973).

a macromolecular complex (reaction VI). Since the calculations make use of the rectilinear and constant-field approximations, the results for the zonal mode of transport are valid for both electrophoresis and sedimentation and apply qualitatively to gel-permeation chromatography.

As illustrated by the representative results displayed in Figures 7 and 8, both moving-boundary and zone patterns for irreversible isomerization, dimerization, and dissociation into subunits typically show two well-resolved

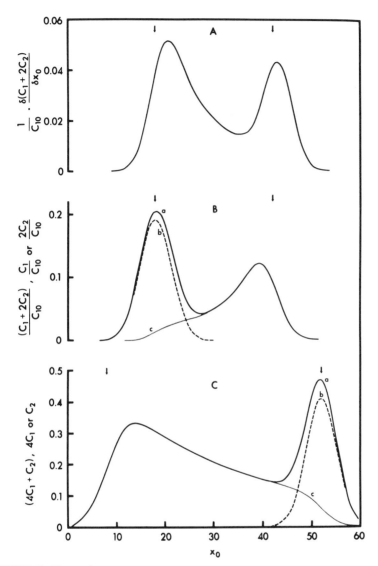

FIGURE 7. Theoretical transport patterns. A—Analytical sedimentation pattern for irreversible dimerization (Reaction V); $v_2 > v_1$; sedimentation is from right to left. B—Zone sedimentation (or electrophoretic) pattern for irreversible dimerization (Reaction V), sedimentation is from left to right: (a) total concentration $(C_1 + 2C_2)/C_{10}$; (b), concentration of monomer C_1/C_{10}; and (c), concentration of dimer $2C_2/C_{10}$. C—Zone sedimentation (or electrophoretic) pattern for irreversible dissociation into four identical subunits (Reaction IV with $m = 4$); $v_1 > v_2$; sedimentation is from left to right: (a) total concentration $4C_1 + C_2$; (b) concentration of tetramer $4C_1$; and (c) concentration of subunit C_2; essentially the same pattern is shown for dissociation into halfmers (Reaction IV with $m = 2$). In each case the macromolecule exists entirely as reactant at the start of sedimentation. From Cann and Oates (1973).

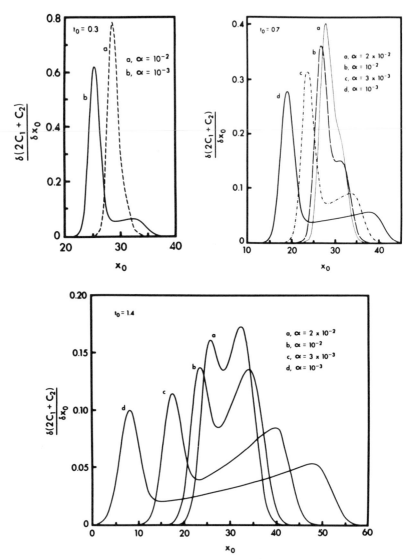

FIGURE 8. Theoretical analytical sedimentation patterns for irreversible dissociation of a macromolecule into halfmers (Reaction IV with $m = 2$), $v_1 > v_2$. Sedimentation is from right to left.

peaks for half-times of reaction ranging from about 0.3 to 2.5 times the duration of the transport experiment, depending upon the difference in velocity between product and reactant. The greater the difference, the better is the resolution for longer half-times. There is no indication of single skewed peaks with long trailing or leading edges as long as the reaction does not go to completion during the course of the experiment. It is characteristic of the patterns (see, e.g., Figures 7B and 7C) that the peak corresponding to a mixture of reactant and some product is sharp, while the one corresponding largely to product is broad and skewed in the direction from whence it was formed. The product peak grows with time at the expense of reactant (Figure 8), and the pattern never reaches the baseline between the peaks. In other words, the two peaks constitute a single reaction boundary or zone. Rate constants can be estimated with a reasonable accuracy from analytical sedimentation patterns. In the case of isomerization or dissociation into subunits, but not dimerization, rate constants estimated from zone patterns are sufficiently accurate to be useful in pilot experiments. These new insights hold promise for sedimentation studies on material such as virus-coat proteins which often tend to be unstable and on the dissociation into subunits of enzyme complexes such as pigeon liver fatty acid synthetase (Kumar *et al.*, 1970; 1972, Kumar and Porter, 1971).

Our calculations for dissociation of a complex into its macromolecular constituents (reaction VI, $k'_2/k'_1 \geq 0$) were designed to simulate the dissociation of either a large-enzyme complex or ribosome–enzyme complex, C, to give a residual complex or ribosome A stripped of a particular enzyme molecule B which now exists in free solution. The question here is "Will the profile of enzymatic activity along the centrifuge tube be bimodal or will it show a single peak which trails?" We proceed by assuming that (1) the driven velocity of B is less than A which, in turn, is the same as C, *i.e.*, $v_1 = v_2 > v_3$; (2) the molecular weight of B is only 10% that of C; and (3) the enzymatic activity per mole of C is the same as the activity per mole of B. The last two assumptions dictate that theoretical zone sedimentation (or electrophoretic) patterns be plots of $C_1 + 0.9C_2 + 0.1C_3$ (*i.e.*, total amount of material) or $C_1 + C_3$ (*i.e.*, enzymatic activity) vs. position. As illustrated by the representative zone sedimentation patterns displayed in Figure 9, both the profile of total material and of enzymatic activity are typically bimodal. The faster-migrating peak corresponds to a mixture of C, A, and B in the pattern of total material and to a mixture of C and B in the pattern of enzymatic activity, while in both cases the slower peak is comprised largely of B. Single skewed peaks with trailing edges are never predicted for patterns of total material, and only if the reaction goes to completion during the course of sedimentation will the profile of activity show a single peak which is skewed forward.

Finally, we note that at least some of the bimodal patterns presented in Figures 7–9 could, in practice, be misinterpreted as indicative of a mixture of two stable, noninteracting macromolecules. Once again appeal must be made to fractionation. In the fractionation test, as applied to zone patterns,

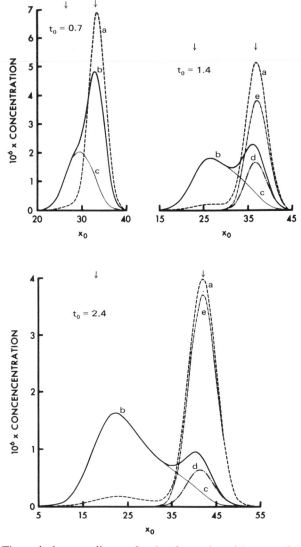

FIGURE 9. Theoretical zone sedimentation (or electrophoretic) pattern for dissociation of a complex (Reaction VI); $v_1 = v_2 > v_3$; sedimentation is from left to right; (a) $C_1 + 0.9C_2 + 0.1C_3$; (b) $C_1 + C_3$; (c) C_3; curve d, C_1; and curve e, C_2. Here $\alpha = 10^{-2}$, $k_2/k_1 = 2 \times 10^5$ M^{-1}, $C_{10} = 10^{-5}$ M, pure complex at the start of sedimentation. From Cann and Oates (1973).

the material in each peak is isolated and analyzed as in the original separation. For reactions of the sort under consideration, the fraction comprised of product will migrate as a single peak while the one rich in reactant will exhibit two peaks like the unfractionated material. For heterogeneity, both fractions will run true. The fractionation test has been applied to the zone sedimentation of enzymatically active complexes between aminoacyl transferase I and proteinaceous cytoplasmic particles (Shelton *et al.*, 1970) and between aminoacyl-tRNA synthetases and ribosomes (Roberts, 1972). The results revealed that the two peaks in the pattern of enzymatic activity do not correspond to isozymes with different molecular weights, but instead constitute a bimodal reaction zone arising from dissociation of the complex during the course of sedimentation.

IV. DISCUSSION

Central to any discussion of the mass transport of interacting macromolecules is the concept of bimodal reaction boundaries or zones, and a physical explanation of resolution into two peaks adds a new level of understanding. Consider, for example, the zone sedimentation of a ligand-mediated dimerizing system for which equilibration is instantaneous (unbroken curve in Figure 3). Because of the coupling between reequilibration during differential sedimentation of macromonomer and dimer and the transport of ligand bound into dimer, the back half of the sedimenting zone is almost depleted of ligand. Consequently, the remaining monomer in this region is deprived of the dimerization-mediating small molecule and thus lags behind to form a second peak. At the same time, the fast peak is enriched in ligand to an extent which more than compensates for dilution of the macromolecule in the spreading zone. As a result, the macromolecule in this peak is actually more highly dimerized than in the starting zone. The two peaks never completely resolve, however, because the association reaction is reversible. A rather different mechanism operates in the case of irreversibly reacting systems in which, for example, the product has a smaller sedimentation velocity than the reactant (Figure 7C for dissociation of a macromolecule into subunits). As the zone of reactant departs from the meniscus and begins to sediment down the centrifuge tube, the reaction is occurring at its maximum rate determined by the initial concentration of reactant. Because of the difference in sedimentation velocity between product and reactant, the product lags behind the advancing peak of reactant and would form a long, trailing plateau if the reaction were to proceed at a constant rate. But, in fact, the rate of reaction decreases progressively with time as reactant is consumed. Since the amount of product formed per unit time decreases as differential sedimentation of product and reactant proceeds, the concentration profile of product must pass through a maximum and skew forward to blend into the peak of reactant. For the special case in which the rate of sedimentation of the product is negligibly small compared to reactant, the maximum in its concentration profile will be located at the meniscus.

Since the generation of bimodal boundaries and zones is not unique for any one reaction mechanism, it is imperative that appeal be made to the combined application of velocity sedimentation with one or more other physical methods in order to establish the exact nature of the interaction. An example is the combined application of analytical sedimentation, Archibald molecular-weight determinations and rapid kinetic measurements to the reversible hexamer–dodecamer reaction of New England lobster hemocyanin (Morimoto and Kegeles, 1971; Tai and Kegeles, 1971; Kegeles and Tai, 1973).

Finally, it is anticipated that the new insights provided by these theoretical advances will find application to a variety of biochemical reactions such as the interaction of enzymes with cofactors and allosteric affectors. Recently, Eisinger and Blumberg (1973) have made a practical application of the mass transport of interacting systems to the determination of the equilibrium constants for the binding of complementary tetranucleotide to tRNA and proflavine to chymotrypsin from acrylamide gel electrophoresis. This work demonstrates how the results of computer-simulated transport can be fitted by semiempirical relationships which show the dependence of the equilibrium constant on various experimental parameters. An alternative procedure advocated by Gilbert and Gilbert (1973) utilizes statistical methods to match computed transport patterns to experimental ones. These investigators took a definitive step in this direction by deriving the equilibrium constants for the self-association of β-lactoglobulin A from analytical sedimentation patterns by the process of computer simulation, comparison with experiment, and iteration. Theirs is an accomplishment which makes clear how velocity sedimentation in conjunction with computer simulation can be truly complementary to sedimentation equilibrium for determining the stoichiometry and equilibrium constants of associating proteins. It is timely, therefore, that Weirich et al. (1973) have explored various ways of combining velocity and equilibrium sedimentation data on self-associating systems. Another interesting development in a somewhat different vein is the introduction of intervent dilution chromatography for separation of strongly interacting macromolecules (Kirkegaard and Agee, 1973). The power of this method, which holds promise for the rapid purification of enzymes, is witnessed by the separation of ribosomal proteins from ribosomal RNA.

V. REFERENCES

Ackers, G. K. (1969) *Adv. Protein Chem.* **24**: 343.
Anderson, P. M. and Marvin, S. V. (1970) *Biochemistry* **9**: 171.
Cann, J. R. (1966) *Biochemistry* **5**: 1108.
Cann, J. R. (1970) *Interacting Macromolecules. The Theory and Practice of Their Electrophoresis, Ultracentrifugation and Chromatography*, Chapter 4, Academic Press, New York.
Cann, J. R. (1973) *Biophys. Chem.* **1**: 1.
Cann, J. R. and Goad, W. B. (1965) *J. Biol. Chem.* **240**: 1162.
Cann, J. R. and Goad, W. B. (1970) *Science* **170**: 441.
Cann, J. R. and Goad, W. B. (1972) *Arch. Biochem. Biophys.* **153**: 603.
Cann J. R. and Kegeles, G. (1974) *Biochemistry* **13**: 1868.
Cann, J. R. and Oates, D. C. (1973) *Biochemistry* **12**: 1112.
Eisinger, J. and Blumberg, W. E. (1973) *Biochemistry* **12**: 3648.
Field, E. O. and O'Brien, J. R. P. (1955) *Biochem.* **60**: 656.
Field, E. O. and Ogston, A. G. (1955) *Biochem. J.* **60**: 661.
Gilbert, G. A. (1955) *Discuss. Faraday Soc.* **20**: 68.
Gilbert, G. A. (1959) *Proc. Roy. Soc. Ser. A* **250**: 377.
Gilbert, L. M. and Gilbert, G. A. (1973) *Meth. Enzymol. D* **27**: 273.
Gilbert, W. and Müller-Hill, B. (1966) *Proc. Natl. Acad. Sci. U.S.A.* **56**: 1891.
Goad, W. B. (1970) in: *Interacting Macromolecules. The Theory and Practice of Their Electrophoresis, Ultracentrifugation, and Chromatography*, J. R. Cann, Chapter 5, Academic Press, New York.
Kegeles, G. and Tai, M. (1973) *Biophys. Chem.* **1**: 46.
Kirkegaard, L. and Agee, C. C. (1973) *Proc. Natl. Acad. Sci. U.S.A.* **70**: 2424.
Kumar, S. and Porter, J. W. (1971) *J. Biol. Chem.* **246**: 7780.
Kumar, S., Dorsey, J. A., Muesing, R. A., and Porter, J. W. (1970) *J. Biol. Chem.* **245**: 4732.
Kumar, S., Muesing, R. A., and Porter, J. W. (1972) *J. Biol. Chem.* **247**: 4749.
Martin, R. G. and Ames, B. N. (1961) *J. Biol. Chem.* **236**: 1372.
Morimoto, K. and Kegeles, G. (1971) *Arch. Biochem. Biophys.* **142**: 247.
Roberts, W. K. (1972) personal communication.
Shelton, E., Kuff, E. L., Maxwell, E. S., and Harrington, J. T. (1970) *J. Cell. Biol.* **45**: 1.
Tai, M.-S. and Kegeles, G. (1971) *Arch. Biochem. Biophys.* **142**: 258.
Weirich, C. A., Adams, E. T., Jr., and Barlow, G. H. (1973) *Biophys. Chem.* **1**: 35.
Weisenberg, R. C. and Timasheff, S. N. (1970) *Biochemistry* **9**: 4110.
Zimmerman, J. K. and Ackers, G. K. (1971) *J. Biol. Chem.* **246**: 1078.
Zimmerman, J. K., Cox, D. S., and Ackers, G. K. (1971) *J. Biol. Chem.* **246**: 4242.

TRANS ELECTROPHORESIS
2

NICHOLAS CATSIMPOOLAS

I. INTRODUCTION

A. Principle

Transient-state* (TRANS) electrophoresis is a method which utilizes repetitive electro-optical scanning of the separation path—in the presence of the electric field—for the kinetic monitoring and computation of the distribution of charged species subjected to various forms of electrophoresis. The separation is performed in a quartz column which bears electrode reservoirs and which moves perpendicularly to a slit of light of adjustable wavelength (Figure 1). The interaction of the light beam with the separated, charged species causes a fluctuation of its intensity (measured as absorbance or percent transmission)

---

* "Transient" refers only to the state of the concentration distribution of a charged species (e.g., protein) subjected to electrophoresis. The "steady state," according to the above definition, is achieved when the first four central moments and the area of the peak remain constant with time.

NICHOLAS CATSIMPOOLAS, Biophysics Laboratory, Department of Nutrition and Food Science, Massachusetts Institute of Technology, Cambridge, Massachusetts 02139. This work was presented at the "Symposium on Recent Developments in Research Methods and Instrumentation," October 22–24, 1974, National Institutes of Health, Bethesda, Maryland and at the 168th American Chemical Society National Meeting Symposium on "Contemporary Protein Separation Methods," September 1974, Atlantic City, New Jersey.

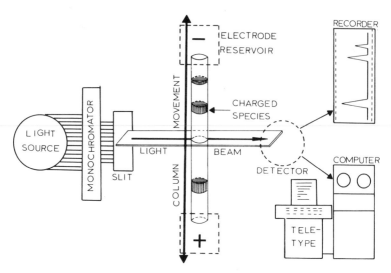

FIGURE 1. Schematic diagram illustrating the basic instrumental aspects of TRANS electrophoretic analysis.

which is detected by a photomultiplier. The current produced by the photocell is converted electronically to an analog voltage by a photometer. The analog signal can be recorded on a strip-chart recorder and also it can be digitized by an analog-to-digital converter and processed by a computer. Repetitive scanning of the separation path produces a series of electropherograms as a function of time which can be utilized in obtaining information relevant to the type of electrophoresis employed for the analysis. A necessary condition to be met is that the charged species of interest (e.g., small molecules, macromolecules, bioparticles, and cells) be detected by the particular wavelength of light used. The interaction of light with the charged species can assume the form of absorbance, fluorescence, or scattering. For quantitative results, a linear relationship between the species concentration and the measured property is required.

B. History of *in Situ* Electro-Optical Scanning

Electro-optical systems using light absorption and a photomultiplier scanning assembly were originally used in ultracentrifugation (Hanlon *et al.*, 1962) and in gel chromatography (Brumbaugh and Ackers, 1968). However, the operational setup in terms of instrumentation and particular methodological problems is significantly different for electrophoresis systems. The most serious attempt at *in situ* electro-optical scanning in electrophoresis was made by Hjertén (1967) with the development of the free-zone electrophoresis apparatus. The method was restricted to separations in free solution and most of the work was centered on measuring electrophoretic mobility by recording the peak position as a function of time. Four other instrumental systems (Olivera *et al.*, 1964; Ressler, 1967; Hochstrasser *et al.*, 1967; Johnsson *et al.*, 1973) were also confined to simple demonstrations of measuring mobility and/or obtaining a graphical record of the separation. Numerous other densitometric techniques have appeared in the scientific literature (Loening, 1967; Dravid *et al.*, 1969; Watkin and Miller, 1970; Fawcett, 1970; Radola and Delincée, 1971; Borris and Aronson, 1969; Easton *et al.*, 1971; Brakke *et al.*, 1968), but these cannot be used for transient electrophoretic analysis because of the absence of the electric field during scanning.

Based on earlier instrumental developments by the author (see Catsimpoolas, 1973a), the transient-state electrophoretic technique was first applied to transient-state isoelectric focusing (TRANS-IF) experiments (Catsimpoolas, 1973*b,c,d,e*; Catsimpoolas and Griffith, 1973; Catsimpoolas *et al.*, 1974*a,b*). The advantages of coupling the *in situ* scanning instrument to an on-line digital data acquisition and processing system were sufficiently demonstrated. A description of the capabilities and potential future applications of TRANS electrophoresis will be presented below.

C. Types of Electrophoresis and Supporting Media

The information that can be obtained by TRANS electrophoresis depends on the type of the method utilized to achieve separation. In principle (Catsimpoolas, 1973*f*), we can distinguish three types of electrophoresis: continuous-pH zone electrophoresis (CZE), isoelectric focusing (IF), and multiphasic zone electrophoresis (MZE). An important subcategory of MZE is isotachophoresis (ITP).

In CZE, the pH of the buffer in both the electrolyte reservoirs and the separation path should remain constant during electrophoresis (Figure 2). The sample is applied as a thin starting zone and the separation depends on differences in electrophoretic mobility of the charged species at the pH of the buffer employed. Although the system is simple, resolution is limited by practical inability to achieve an ultrathin starting zone and by band spreading due to diffusion during migration. A large number of buffers (>4269) for

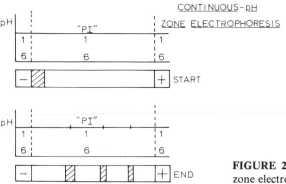

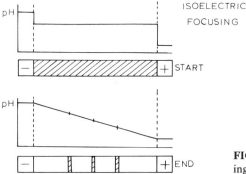

FIGURE 2. Stages of continuous-pH zone electrophoresis (CZE).

FIGURE 3. Stages of isoelectric focusing (IF).

CZE can be prepared from constituents 1 and 6 (PI phase)* as described in the *Multiphasic Buffer Systems Catalogue* (Jovin *et al.*, 1970). These buffers are specifically described in terms of pH, ionic strength, conductance, mobility and degree of ionization of constituent electrolyte components, and other useful parameters.

In IF (Catsimpoolas, 1973*d*), the sample in mixture with special amphoteric compounds called "carrier ampholytes" is placed between strongly acidic and strongly basic electrolyte reservoirs positioned at the positive and negative electrodes, respectively (Figure 3). When the electric field

---

* Nomenclature of some constituents and phases used in MZE: ALPHA, upper buffer phase prior to electrophoresis; BETA, stacking phase prior to electrophoresis; Constituent 1, trailing ion of the stack; Constituent 2, leading ion of the stack; Constituent 3, ion of the separation phase prior to electrophoresis; Constituent 6, counter ion common to all phases; GAMMA, separation phase prior to electrophoresis; LAMBDA, separation phase after migration of the leading ion of the stack into the GAMMA phase; PI, operative separation phase; and ZETA, operative stacking phase.

is applied, the carrier ampholytes form a stable pH gradient in which the amphoteric charged species to be separated migrate until they reach the pH corresponding to their isoelectric point (pI). Since the net electric charge of the molecules at pI is zero, there is no further migration in the electric field and, thus, compounds become concentrated or "focused" at different parts of the pH gradient depending on their respective pI values. The IF method offers a very high resolution tool for the separation of proteins differing as little as 0.01 pH unit in their pI values.

In MZE (Jovin, 1973 and references therein; Ornstein, 1964) a system of discontinuous buffer phases is utilized to achieve a "stacking" of the components into thin zones followed by an "unstacking" in the resolving phase (Figure 4). Charged species are separated according to their electrophoretic mobility in the PI phase, which is generated during electrophoresis. In the

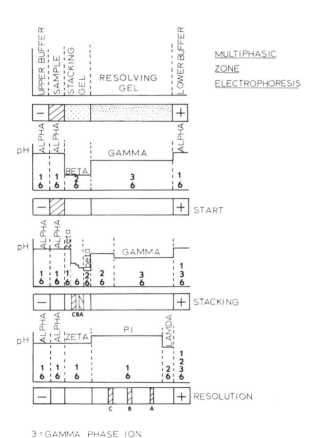

3 = GAMMA PHASE ION

FIGURE 4. Stages of multiphasic zone electrophoresis (MZE).

stacking phase, the components are concentrated into a thin zone because their relative constituent mobilities fall within the range specified by the relative constituent mobility of constituent 1 in the phase ZETA and that of constituent 2 in phase BETA. The steady-state stacking phase is also called isotachophoresis (ITP) (Figure 5) and can be carried out independently of the resolution phase of MZE (Catsimpoolas, 1973*f*). In general, constituent 1 must be a monovalent weak electrolyte, but constituent 2 can be an ion, or a monovalent or divalent weak electrolyte. Constituent 6 is common to all phases and can be either an ion or a monovalent weak electrolyte. The ALPHA phase (upper buffer) should normally be identical to the ZETA phase. In MZE, when the stack reaches the interface between the stacking and resolution buffers, a programed pH change takes place such that the relative constituent mobility of constituent 1 (in the phase PI) is greater in magnitude-than that of the sample components, thus, migrating ahead of them in the PI phase. This results in a system that resembles CZE, but is not identical to CZE because of the presence of a moving boundary (PI/LAMBDA) which can often be visualized by the presence of a tracking dye such as bromophenol blue.

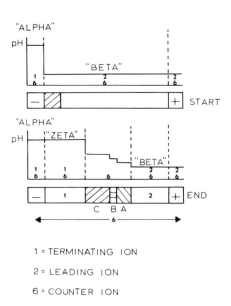

FIGURE 5. Stages of isotachophoresis (ITP) or steady-state stacking (SSS).

In addition to the choice of electrolyte system (CZE, IF, MZE, ITP), information obtained by TRANS electrophoresis depends also on the supporting medium, which can be either a density gradient or a gel (such as agarose, Sephadex, or polyacrylamide). Electrophoretic migration in a density gradient is only influenced by the viscosity gradient and possible interactions of the densor (e.g., sucrose, Ficoll) with the species of interest. On the other hand, migration in "sieving" gels (e.g., polyacrylamide) is subject to "retardation" depending on the structure of the gel and the size and shape of the macromolecule. Thus, additional parameters can be obtained pertaining not only to charge—as expected in an electrophoretic process—but also to molecular size (Chrambach and Rodbard, 1971; Rodbard and Chrambach, 1971). These phenomena will be examined in more detail in the following text.

D. Anatomy of the TRANS Electrophoresis System

The TRANS electrophoresis system represents an integrated approach to electrophoresis and at present consists of the following parts:

1. *In situ* electro-optical scanning in the presence of the electric field.
2. Computerization of data acquisition and processing used extensively in other separation techniques, e.g., ultracentrifugation and chromatography.

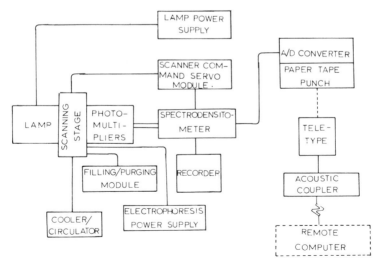

FIGURE 6. Block diagram of the TRANS electrophoresis instrument.

3. Peak shape analysis, the usefulness of which is adequately described by Grushka in Chapter 6 of this volume.
4. Adoption of Jovin's system for the design and analysis of discontinuous buffer systems with a digital computer (Jovin, 1973; Jovin *et al.*, 1970).
5. Adoption of the Rodbard–Chrambach system (Rodbard and Chrambach, 1974; Rodbard *et al.*, 1974) for the mathematical and statistical analysis of polyacrylamide gel electrophoresis (PAGE) data and for the optimization of resolution in analytical and preparative PAGE.
6. Kinetic analysis of isoelectric focusing (TRANS-IF) described by the author and co-workers.

The above systems will be expanded in the future to include kinetic electrophoretic methods for studying protein interactions.

II. INSTRUMENTAL ASPECTS

A. The Electro-Optical Unit

Four prototype instrumental assemblies (Catsimpoolas, 1971*a,b,c*; Catsimpoolas *et al.*, 1975) have been used for TRANS electrophoresis experiments. Since one of those is in certain respects more advanced than the others, it will be described here in more detail. A schematic diagram of this instrument is shown in Figure 6 and a photograph of it is shown in Figure 7.

A 200-W xenon–mercury arc lamp and associated power supply (Schoeffel) are used in conjunction with a tandem grating monochromator (Schoeffel GM100D) to produce monochromatic energy in the 200–700 nm wavelength range with very low stray-light characteristics (e.g., $1:10^4$ at 220 nm) (Figure 8). The monochromatic light beam is subsequently collimated by a quartz lens, shaped by a variable horizontal slit (25–100 μm), passed through the electrophoresis cells, and is thereafter directed to the sample and reference photomultipliers. The light beam is divided to simultaneously illuminate the sample and reference cells. Each beam is detected separately by its photomultiplier and the log of the ratio of both signals i.e., the linear absorbance, is provided electronically (Schoeffel Model SD 3000 Spectrodensitometer). The ratioing of the signals eliminates errors due to possible fluctuations of light intensity and photomultiplier high voltage. A knob balance control and zero meter (provided on the panel) permit the matching of spectral sensitivity levels of the photomultipliers in any desired spectral range. This adjustment makes the electrical outputs equal (signal

FIGURE 7. Photograph of the TRANS electrophoresis instrument.

ratio of 1:1), which is equivalent to zero optical density (O.D.) at the beginning of the scan, with the sample beam in a "neutral" area of the media to be measured. A panel meter (0.0–3.0 O.D. units) is also provided with a "zero" control knob. Seven O.D. full-scale ranges (0.1, 0.2, 0.4, 1.0, 2.0,

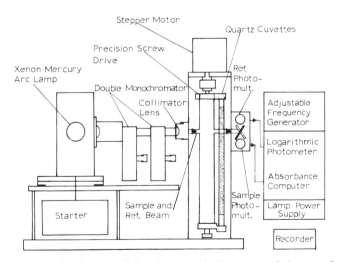

FIGURE 8. Schematic diagram of the electro-optical system and the scanning stage.

4.0, and 10.0) can be selected on the front panel. The photometer supplies an electrical output of 1 V per 1 O.D. unit for computer processing and 100 mV full scale (at any O.D. range) for operating a recorder. Single (sample photo-multiplier only) or ratio recording of absorbance and scattering is available in the instrument.

B. The Scanning Stage Module

This part of the instrument provides dark housing for the removable electrophoresis cassette and vertical (linear) transport stage by means of a reversible stepping motor attached to a lead screw (Figure 8). The stepping motor and adjustable frequency generator provide six reversible scanning speeds of 5, 10, 20, 40, 80, and 160 sec/cm. Upper and lower microswitch stops make possible manual actuation and automatic stopping of the scanner at predetermined column intervals.

C. The Scanner Control Servo Unit

The linear transport stage and associated instrumentation are automatic-ally controlled by the "scanner control servo unit." This is accomplished as follows: A 2.5-kΩ "scanner pickoff" linear displacement potentiometer is mechanically fixed to the moving stage. The potentiometer is driven from an appropriately scaled voltage source; this results in the potentiometer output voltage being numerically equivalent to the position of the scanner in centi-meters to 0.05% accuracy. This potential is impedance transformed and fed to a digital panel meter (Newport 2000) which provides a digital readout of the scanner position in centimeters. The signal is also supplied to two ampli-fiers which are in a double-limit comparator configuration. Two 10-kΩ ten-turn potentiometers serve to allow setting up "high" and "low" scanner travel-limit positions and are calibrated directly in centimeters to three places. The comparator continuously compares the limit potentiometer settings with the pickoff signal and issues motor-reversing commands to the memory/switching element accordingly. The memory/switching element is essentially equivalent to a set–reset flip-flop, but it features versatile switching capability and high noise immunity. One set of relay contacts function as a latch and provide the "memory" function. A second set of contacts sources the reversing information to the motor. Another set of contacts automatically operates the "sweep" and "reset" functions of an x–y recorder and also commands the paper tape punch to issue a leader (blank space on tape) space on the tape. A final set of contacts provides for the motor to run at a higher speed in the "down" scan mode. All the functions of the scanner command servo may be overridden (and performed manually) through panel switches.

The practical implications of the scanner control servo unit are several. The scanning interval (in centimeters) can be selected digitally and be set at any part of the column by means of two control knobs (high and low position). A digital display of column position with respect to the light beam is available at all times during scanning. The recorder and paper tape punch are actuated automatically during the entire electrophoretic experiment. Finally, the scanner can be commanded to run automatically at different speeds in the "up" or "down" mode of travel. This allows collecting data either in both directions of travel, or preferentially at normal speed in the "up" mode accompanied by fast reversal in the "down" mode without data collection.

D. The Electrophoresis Cell Cassette

This part of the assembly is removable and accommodates two quartz columns (Amersil, Suprasil T-20) which are used for electrophoresis, isotachophoresis, or isoelectric focusing experiments (Figure 9). The dimensions

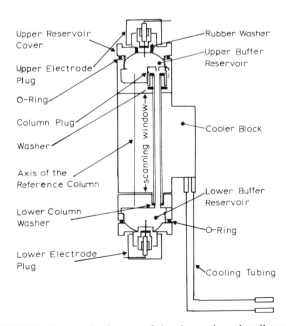

FIGURE 9. Schematic diagram of the electrophoresis cell cassette.

of the columns are: 6.0 mm i.d., 8.0 mm o.d., and 142 mm length. Only one column carries the sample for analysis, the other acts as reference background in balancing the photometer output. In density gradient experiments the bottom of the column is covered either with a semipermeable membrane or with a polyacrylamide gel plug. The cassette also accommodates the upper and lower flow-through electrolyte reservoirs with removable platinum electrodes and a flow-through cooler block for circulating cooling liquid. An additional outlet is provided at the top of the lower reservoir to allow escape of air bubbles during loading. The total cassette can be easily disassembled for cleaning.

E. The Filling/Purging/Cooling Module

This unit consists of a multichannel Technicon Autoanalyzer proportioning pump, a vacuum pump, Hamilton chemically inert valves, a thermostatted cooler/circulator, and associated tubing (Figure 10). The function of the assembly is to recirculate the buffers into the upper and lower electrolyte

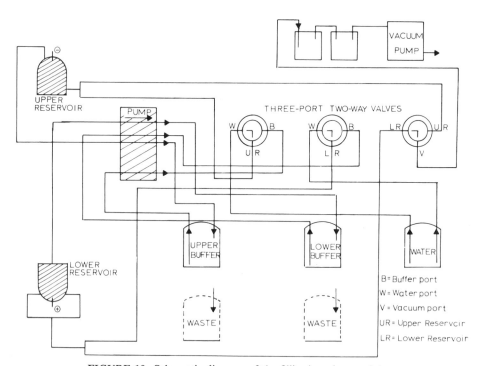

FIGURE 10. Schematic diagram of the filling/purging module.

reservoirs during the electrophoretic run, purge and wash the flow system at the end of the run, and circulate fluid through the cooling block. The buffer glass reservoirs are kept in the bath of the cooler for temperature equilibration.

F. Other Components

A regulated d.c. power supply providing both constant voltage and constant current operation is used for electrophoresis and isoelectric focusing experiments. The photometer output is recorded with an Esterline-Angus Model 2417 TB $x-y$ and $y-t$ recorder automatically controlled by the scanner control servo unit.

G. The Digital Data Acquisition Module

This unit consists of an anolog-to-digital converter with visual display, two-channel analog multiplexer, variable-speed sampler, a paper tape punch, a teletype, and an acoustic coupler. Data can be punched at a switch-selectable rate of 1–20 readings per second. Other provisions include choice of three gain settings, single- or double-channel operation, and single or continuous sampling of the data. The acquired data punched on the paper tape are processed by a remote computer via the teletype acoustic coupler module. In addition a digital multiplexer–logger provides a readout of temperature, voltage, current, and power on a teletype.

H. General Operational Procedure

The upper and lower buffer in glass containers are placed in the constant-temperature bath for thermal equilibration. Cooling fluid also is allowed to flow through the cooling block of the electrophoresis cell cassette. The quartz column, covered at the lower end with a dialysis membrane sheet premoistened with the lower buffer, is placed in the cassette and the density gradient or gel is prepared *in situ*. Concurrent with the above operation, the buffer is pumped into the lower electrolyte reservoir, which is tilted at a slight angle to prevent bubble entrapment under the column. The sample solution is then applied on top of the gradient and carefully layered with the upper buffer, thus filling the remainder of the column. Subsequently, the buffer is pumped into the upper reservoir with the upper chamber cover in position. The upper electrode plug is then inserted carefully to avoid pressure in the column. The electrophoresis power supply is turned on for a few seconds to establish the passage of current. The electrophoresis cassette is placed in the scanning stage module and the lid is closed. The photometer is adjusted to zero at a "neutral" area of the column and a preliminary scan is obtained to establish the baseline before the current is turned on for the experiment. During electrophoresis the buffers are circulating through the upper and lower electrode reservoirs, thus removing undesirable products and air bubbles generated by the platinum electrodes. The column is scanned continuously during the experiment, providing both an analog signal recorded on the recorder chart and digital data punched on the paper tape. At the conclusion of the experiment, the pump is stopped and the buffer in the reservoirs and tubing is removed by vacuum suction. The column is removed, appropriate plugs are inserted, and the total buffer circulation system is washed with water and purged repetitively by means of the filling/purging module.

I. Column-Coating Procedure

The quartz column is coated before use with density gradients to prevent electroendosmosis by a modification of the procedure described by Hjertén (1967). Methylcellulose (Methocel, visc. 8000 cp, Dow Chemical Company) (0.4 g) is dispersed in 30 ml hot (boiling) water and stirred until a lumpfree dispersion is formed. An additional 70 ml of cold (4°C) water is added and stirring is resumed in the cold room until the solution appears clear. Formic acid (7 ml) and then formaldehyde (35 ml) are added, with stirring. The final solution is clarified by filtration and can be stored in the refrigerator for at least 6 months. The quartz tube is washed thoroughly with a detergent solution, hot and cold tap water, distilled water, and then dried. The methylcellulose solution is drawn into the tube by suction and after 5 min is allowed to run out slowly. Subsequently, the tube is dried in a 120°C oven for 40 min. The coating and drying procedures are repeated once more. Care should be exercised to avoid air-bubble adherence to the tube wall and to keep the tube in the vertical position during coating.

III. DATA PROCESSING

A. Slope Analysis

The analog signal produced by the photometer is digitized by an analog-to-digital converter at preselected time intervals. Since the mechanical scanning speed is constant, the digitized values represent the photometer signal

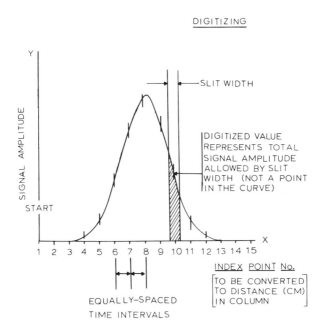

DIGITIZING

FIGURE 11. Diagram of the process of digitizing the absorbance profile of a distribution.

amplitude at equidistant points (i.e., center column of the window width formed by the slit) (Figure 11). The number of data points obtained per peak j depends on the velocity v (cm/sec) of the scanner, the digitizing rate g (data points/sec), and the total width of the peak profile at the baseline w_t (cm), so that

$$j = gw_t/v$$

To a first approximation, the narrower the slit width s, the more accurately the peak profile will be recorded. However, the noise is also expected to increase because the larger the slit width the more smoothing of the data occurs by averaging of the light intensity over a larger window. In practice, the slit width should be approximately equal to the ratio w_t/j. Depending on baseline noise and peak shape (Catsimpoolas and Griffith, 1973) the desired value of j can be set between 10 and 100.

Noisy data can be smoothed digitally by using a five-point third-order polynomial fitting technique:

$$d_{sm} = (-3d_{i-2} + 12d_{i-1} + 17d_i + 12d_{i+1} - 3d_{i+2})/35$$

where d_{sm} is the smoothed ordinate and $d_{i \pm n}$ the observed values of the moving box. Slope analysis (i.e., computation of the first and second derivative) can be carried out by determination of the orthogonal linear (L_i) and quadratic (Q_i) components of the curve (Figure 12):

$$L_i = (-2d_{i-2} - d_{i-1} + 0d_i + d_{i+1} + 2d_{i+2})/10$$
$$Q_i = (2d_{i-2} - d_{i-1} - 2d_i - d_{i+1} + 2d_{i+2})/7$$

The L_i and Q_i curves are useful in detection of the start and end of peak, overlapping peaks (see Chapter 6), and the inflection point of the distribution.

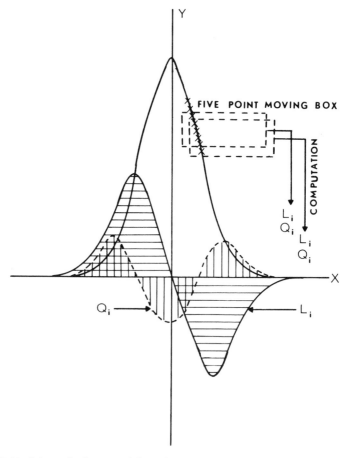

FIGURE 12. Schematic diagram of the orthogonal linear (L_i) and quadratic components (Q_i) of a Gaussian distribution.

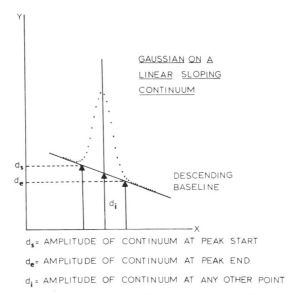

FIGURE 13. Principle of baseline correction by assuming a linear-baseline sloping continuum under the curve.

Correction of the distribution for ascending or descending baseline is subsequently carried out by linear interpolation assuming a linear continuum under the curve (Figure 13) using the equation

$$d_{corr} = d_{obs} - [(d_e - d_s)/(x_e - x_s)]x_{obs} - [(d_s x_e - d_e x_s)/(x_e - x_s)]$$

where d_{corr} is the corrected ordinate, d_{obs} the observed ordinate, x_{obs} the corresponding coordinate, d_s and d_e the ordinates at peak start and end, and x_s and x_e the corresponding coordinates.

B. Moment Analysis

After baseline correction, the first four statistical moments of the peak from an arbitrary origin are computed as follows:

$$m'n = d_i x_i^n / d_i$$

where $n = 1, 2, 3, 4$; d_i is the ordinate at the ith interval; and x_i is the abscissa at the ith interval. The corrected central moments are then computed by

$$m_2 = m'_2 - (m'_1)^2$$
$$m_3 = m'_3 - 3m'_1 m'_2 + 2(m'_1)^3$$
$$m_4 = m'_4 - 4m'_1 m'_3 + 6(m'_1)^2 m'_2 - 3(m'_1)^4$$

and the skewness (S) and excess (E) are

$$S = m_3/\sqrt{m_2}^{3/2}$$
$$E = (m_4/m_2^2) - 3$$

Characterization of a peak by its individual moments can be expressed as follows: The zeroth moment is its area; the first moment (from an arbitrary origin) is the position of the center of gravity (mean) of the concentration profile; the second central moment is the peak variance (square of the standard deviation, σ); the third central moment shows the magnitude and direction of asymmetry; and the fourth central moment is a measure of peak flatness as compared to the Gaussian distribution. The dimensionless coefficients S and E are functions of the moments and reflect asymmetry and flattening relative to the Gaussian form (Figure 14).

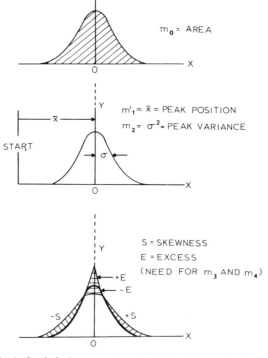

FIGURE 14. Statistical moments and their significance in electrophoresis.

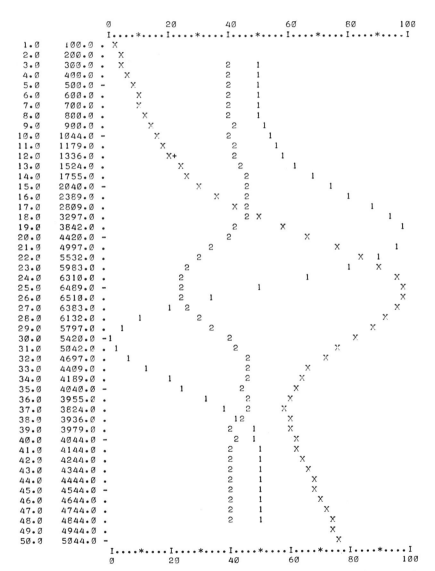

FIGURE 15. Sample computer output showing the observed values of a Gaussian distribution on an ascending baseline and the plot of the L_i and Q_i components.

```
HORIZONTAL Y RANGE-        Ø TO     4000.00 IN INCREMENTS OF    80.0000
VERTICAL    X RANGE-        8 TO         41 IN INCREMENTS OF         1


                      Ø        20       40        60       80      100
                      I....*....I....*....I....*....I....*....I....*....I
       8.0       .Ø .X
       9.0       Ø*.X
      10.0      41.3 .+Ø
      11.0      75.0 . X
      12.0     130.7 .  X
      13.0     217.3 .    X
      14.0     347.0 .      X
      15.0     530.7 .         X
      16.0    ·778.3 .           X
      17.0    1097.0 .              X
      18.0    1483.7 .                 X
      19.0    1927.3 .                    X
      20.0    2404.0 .                       X
      21.0    2879.7 .                          X
      22.0    3313.3 .                            X
      23.0    3663.0 .                              X
      24.0    3888.7 .                                X
      25.0    3966.3 .                                 X
      26.0    3886.0 .                                X
      27.0    3657.7 .·                             X
      28.0    3305.3 .                           X
      29.0    2869.0 .                         X
      30.0    2390.7 .                      X
      31.0    1911.3 .·                  X
      32.0    1465.0 .               X
      33.0    1075.7 .            X
      34.0     754.3 .         X
      35.0     504.0 .       X
      36.0     317.7 .     X
      37.0      85.3 . Ø+
      38.0      96.0 . X
      39.0      37.7 .Ø+
      40.0       1.3 .X
      41.0        Ø .X
                      I....*....I....*....I....*....I....*....I....*....I
                      Ø        20       40        60       80      100
```

```
DATA BLOCK      1
PEAK NO.        1
VARIATION NO.   1
```

		OBSERVED	SMOOTHED
X MEAN	=	24.91	24.91
2ND MOMENT ABOUT MEAN	=	23.45	23.48
3RD MOMENT ABOUT MEAN	=	-6.75	-6.76
4TH MOMENT ABOUT MEAN	=	1533.09	1539.50
BETA 1	=	-.06	-.06
BETA 2	=	2.79	2.79
STANDARD DEVIATION	=	4.84	4.85
AREA	=	49100.33	49109.16

FIGURE 16. Sample computer output of the distribution shown in Figure 15 after baseline subtraction and estimation of the moments.

A computer program available both in FORTRAN and BASIC languages is routinely used for processing TRANS electrophoresis data in this laboratory. A sample output showing a Gaussian distribution superimposed on an ascending baseline and its subsequent baseline correction and calculation of the moments is shown in Figures 15 and 16. Other programs are also available for routine statistical analysis. In addition, a computer program for the mathematical and statistical analysis of polyacrylamide gel electrophoresis data has been made available to us by Drs. D. Rodbard and A. Chrambach (Rodbard and Chrambach, 1971; Rodbard et al., 1974).

IV. TRANS-CZE AND TRANS-MZE

A. Preliminary Considerations

Several parameters relevant to continuous-pH zone electrophoresis (CZE) and multiphasic zone electrophoresis (MZE) can be measured by the TRANS electrophoresis method. These are summarized schematically in Figure 17 and include the peak position, velocity, variance, spreading in the presence ($\mathscr{E} > 0$) and absence ($\mathscr{E} = 0$) of the electric field, mobility relative to the front (R_f), boundary displacement (NU), and resolution (R_s). Accurate measurement of the relative mobility (R_f) in polyacrylamide gel leads to the estimation of other methodological parameters and physical constants as depicted in Figure 18. These measurements and their significance in protein separation and characterization have been discussed in detail by Chrambach and Rodbard (1971). This review is concerned mostly with the kinetic aspects of CZE and MZE, i.e., the measurement of peak position and variance as a function of time.

B. Peak Velocity

The peak velocity ($\bar{v}$) in CZE and MZE can be measured from the slope of the plot of peak position ($\bar{x}$) vs. time (t).

$$\bar{v} = d\bar{x}/dt$$

Peak position ($\bar{x}_i$) is measured at different time intervals of electrophoresis in respect to the peak position ($\bar{x}_r$) of a stationary reference mark on the quartz tube (Figure 19). It is not good practice to use the interface between the upper buffer and the supporting medium (e.g., gel or density gradients) as a reference mark because it may be unstable with time. Thus, the position of the interface can also be evaluated. The time at which the peak centroid (m_1') coincides with the position of an infinitely sharp interface is designated

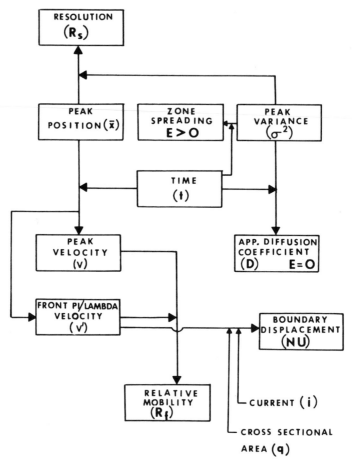

FIGURE 17. Flow diagram of the relationship among various parameters measured in TRANS-CZE or TRANS-MZE.

as t^*. Therefore, the residence time of the peak in the supporting medium is $t - t^*$, where t is the elapsed time after the electric field is applied. Because of experimental difficulties in measuring t^* directly, its value can be obtained by extrapolation and corresponds to the t-axis intercept of the $\bar{x}$ vs. t plot (Figure 20). The values of $\bar{v}$ and t^* are obtained by linear regression:

$$\bar{v} = \frac{n \sum t\bar{x} - (\sum t)(\sum \bar{x})}{n \sum (t^2) - (\sum t)^2}$$

$$t^* = (\sum \bar{x} - \bar{v} \sum t)/n$$

where n is the number of observations.

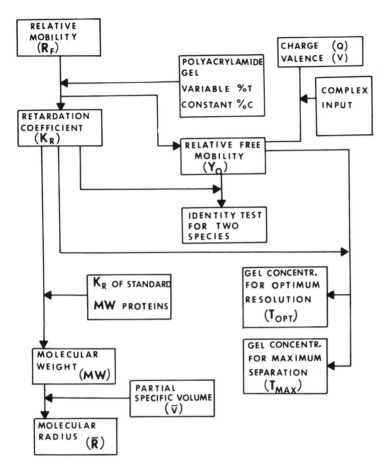

FIGURE 18. Flow diagram of additional parameters that can be measured in poly-acrylamide gel electrophoresis (PAGE).

If the velocity of the front ($\bar{v}$) (PI/LAMBDA boundary in MZE) is measured (e.g., by using a tracking dye) simultaneously with the protein of interest, the R_f value of the latter is obtained from the ratio of the velocities

$$R_f = \bar{v}/\bar{v}'$$

This approach in estimating R_f values of proteins is expected to be more accurate than currently practiced techniques which involve only one measurement of peak positions at the end of the experiment and in the absence of the electric field whereby the centroid of the peak is judged visually. An

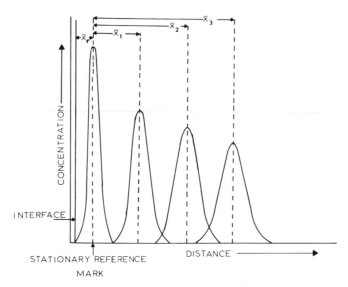

FIGURE 19. Diagram depicting the measurement of peak position as a function of time in TRANS electrophoresis.

example of measuring the R_f value of ovalbumin in a MZE (8%T, 5%C) system is shown in Figure 21.

C. Kinetic Peak-Variance Measurements

The statistical analysis of the second central moment of a peak distribution provides a measure of its variance (σ^2). Changes in peak variance as a function of time can be utilized in the presence of the electric field ($\mathscr{E} > 0$) to determine peak spreading (Figure 22) and in the absence of the electric field ($\mathscr{E} = 0$) to determine the apparent diffusion coefficient D (Figure 23). D can be estimated from the slope of the plot of σ^2 vs. t from

$$D = \frac{1}{2}\frac{d(\sigma^2)}{dt}$$

Peak spreading in the presence of the electric field can be attributed to molecular diffusion, heterogeneity, and imperfections in the supporting medium. Measurement of σ^2 ($\mathscr{E} > 0$) in analytical polyacrylamide gel electrophoresis (PAGE) can provide useful information for the optimization of resolution in preparative PAGE (Rodbard *et al.*, 1974). An example of band spreading ($\mathscr{E} > 0$) of ovalbumin in a CZE (8%T, 5%C) system is shown in Figure 24.

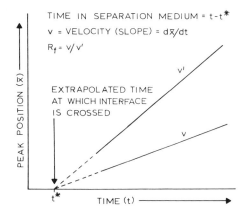

FIGURE 20. Diagram illustrating the measurement of peak velocity and t^*.

Examples of estimation of the apparent diffusion coefficient D will be given in the TRANS-IF section.

D. Resolution

In view of previous considerations concerning resolution in isoelectric focusing (Catsimpoolas, 1973e), we define resolution as

$$R_s = \Delta\bar{x}/(\sigma_A{}^2 + \sigma_B{}^2)^{1/2}$$

where $\Delta\bar{x}$ is peak separation and $\sigma_A{}^2$ and $\sigma_B{}^2$ are the variance of peaks A and B, respectively (Figure 25). Since both $\Delta\bar{x}$ and σ^2 can be measured directly by TRANS electrophoresis, the resolution can be estimated as a function of time.

E. Boundary Displacement

Boundary displacement (Rodbard and Chrambach, 1971), either of the BETA/ZETA or PI/LAMBDA boundaries in MZE, can be measured directly by TRANS electrophoresis because the refractive-index change produces a sharp peak which is detectable with the photomultiplier. The boundary displacement (NU) can be calculated as

$$NU = \bar{v}/iq$$

where $\bar{v}$ is the velocity of the boundary (cm/sec), i is the current (A), q is the cross-sectional area (cm^2), and NU is the boundary displacement (cm^3/C).

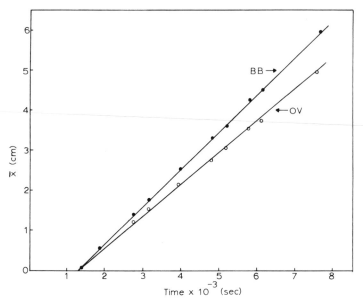

FIGURE 21. Example of measurement of velocity and R_f values of ovalbumin (OV) using bromophenol blue (BB) as a tracking dye.

These measurements can also be utilized for obtaining correction factors for retardation of tracking dues behind the PI|LAMBDA boundaries.

V. TRANS-IF

A. Preliminary Considerations

In analogy to the TRANS-CZE and TRANS-MZE systems, a number of relevant parameters and physical constants can be estimated in transient-state isoelectric focusing (TRANS-IF) (Catsimpoolas, 1973*b,c,d,e*, 1974, 1975*a,b*; Catsimpoolas *et al.*, 1974*a,b*; Weiss *et al.*, 1974), by using moment analysis as outlined in Figure 26. Theory and experimental results concerning such measurements have been reported as follows: resolution and resolving power (Catsimpoolas, 1973*e*), minimal focusing time (Catsimpoolas, 1973*c*; Catsimpoolas *et al.*, 1974*a*), segmental pH gradient and isoelectric point (p*I*) (Catsimpoolas, 1973*c*, 1975*a*), retardation coefficient (Catsimpoolas, 1973*d*), parameters *pE*, *p*, and $dM/d(\text{pH})$ (Catsimpoolas *et al.*, 1974*b*; Catsimpoolas, 1974; Weiss *et al.*, 1974), and instability of peak position, peak area, and segmental pH gradient (Catsimpoolas, 1973*c*).

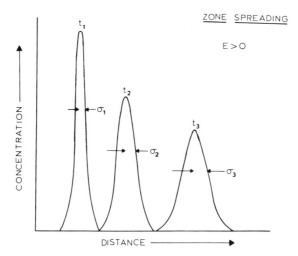

FIGURE 22. Diagram illustrating band spreading in electrophoresis in the presence of the electric field.

B. Minimal Focusing Time

Minimal focusing time (t_{MF}) is defined as the time required for an amphoteric compound to migrate to its pI position. In performing this type of measurement, two general cases can be distinguished: t_{MF} is measured either by using one compound (model ampholyte or protein) or a mixture of

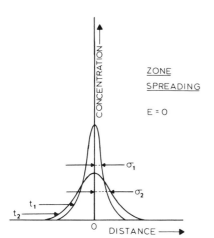

FIGURE 23. Diagram depicting band spreading in electrophoresis by zonal diffusion in the absence of the electric field.

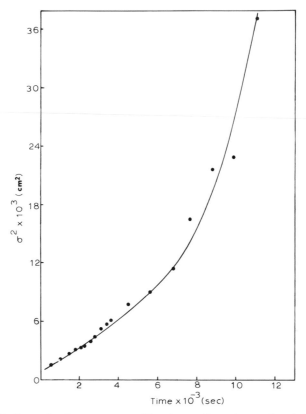

FIGURE 24. Example of measurement of band spreading using ovalbumin in a TRANS-CZE system (8% T, 5% C) in the presence of the electric field.

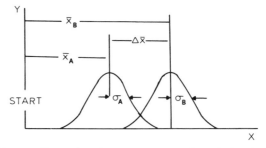

FIGURE 25. Diagram illustrating the measurement of resolution in TRANS electro-phoresis.

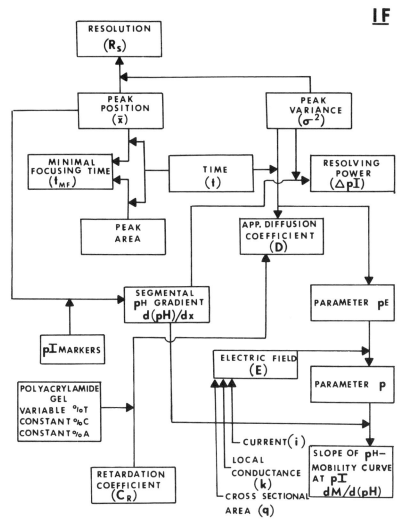

FIGURE 26. Flow diagram of the relationship among various parameters and apparent physical contacts measured in TRANS-IF.

compounds. In the most complex case where a mixture of compounds is used, it is impossible to follow the kinetics of migration of each peak. Therefore, repetitive measurements of the area of each discernible peak at the pI position are made until the steady state is achieved. The peak area should increase with time, attain a maximum value, and remain constant at the steady state. This approach to the measurement of t_{MF} requires that the protein or model ampholytes be loaded in the uniform or gradient mode rather than in the pulse fashion.

When only one compound is used, t_{MF} can be measured by recording the peak position during focusing. Generally, in the pulse mode of loading, a homogeneous sample (one amphoteric species) will migrate as a discernible peak toward the pI position ($\bar{x}_0$), where its electrophoretic mobility will gradually approach zero (Figure 27). In the absence of electro-osmotic flow, or any other condition causing migration of a focused zone toward the electrodes, the t_{MF} can be evaluated in the following manners: (a) by recording the peak position $\bar{x}$ as a function of time until it becomes constant at the steady state $\{f[\bar{x}(t)] = \text{constant}\}$ or (b) by computing the peak velocity $d\bar{x}/dt$ until it reaches zero at the pI position $\{f[d\bar{x}/dt(t)] = 0\}$. In the uniform or gradient mode of loading, the sample forms two discernible peaks mi-

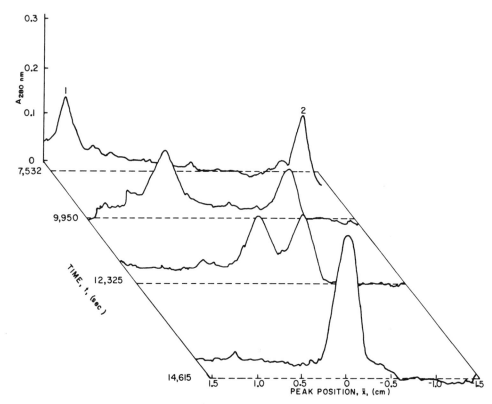

FIGURE 27. TRANS electrophoretic patterns of histidyl–tyrosine peaks migrating toward the pI position (arbitrarily set at $\bar{x} = 0$). Peak 1 migrates away from the negative electrode and peak 2 from the positive electrode.

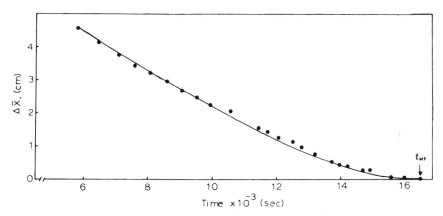

FIGURE 28. Plot of peak position difference ($\Delta \bar{x}$) vs. time for estimation of t_{MF}. Sample: histidyl–tyrosine; 3% Ampholine concentration.

grating away from the two electrodes toward the pI position. In this case, in addition to the above criteria for determining t_{MF}, one can also use the peak position difference ($\Delta \bar{x} = \bar{x}_C - \bar{x}_A$) as a function of time for deriving the same parameter. The notations $\bar{x}_C$ and $\bar{x}_A$ denote the peak positions of the discernible peaks migrating from the cathode and anode of the column toward the pI position, respectively. The distance in the column x, is considered as being increased from the cathode toward the anode. The pI position ($\bar{x}_0$) is reached when $f[\Delta \bar{x}(t)] = 0$. Figures 28 and 29 illustrate two of these experimental approaches, i.e., measurement of $\bar{x}$ or $\Delta \bar{x}$, for the determination of t_{MF} of histidyl–tyrosine and soybean trypsin inhibitor.

The method can be used to evaluate t_{MF} as a function of carrier ampholyte concentration, polyacrylamide gel concentration, viscosity (e.g., in sucrose density gradient IF), electric field strength, and additives (e.g., urea).

C. Segmental pH Gradient and Apparent Isoelectric Point

The segmental pH gradient, $\Delta(\text{pH})/\Delta x$ (cm^{-1}), is measured using two pI markers of closely spaced isoelectric points from

$$\frac{\Delta(\text{pH})}{\Delta x} = \frac{\text{p}I_A - \text{p}I_B}{\bar{x}_A - \bar{x}_B}$$

where pI is the isoelectric point, $\bar{x}$ is the peak position, and subscripts $_A$ and $_B$ denote two pI markers (Figure 30). In using the above equation, it is

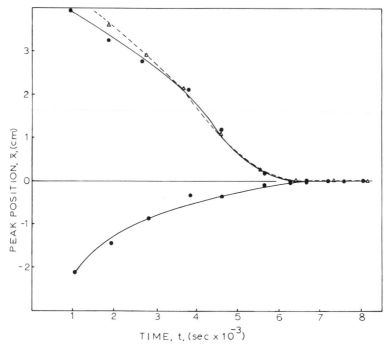

FIGURE 29. Plot of peak position vs. time for measurement of t_{MF}. Sample: soybean trypsin inhibitor; 2% Ampholine concentration. Solid circles represent uniform sample loading and open triangles represent pulse loading.

assumed that species A and B have reached their isoelectric point, and that $\Delta(\text{pH})/\Delta x$ is constant between pI_A and pI_B, where $\bar{x}_A - \bar{x}_B$ represents a small segment of the separation path.

If an "unknown" protein U is included in the segmental pH gradient as described above, its apparent isoelectric point can be calculated by

$$pI_U = pI_A + \left(\frac{\Delta(\text{pH})}{\Delta x}\right)(\bar{x}_A - \bar{x}_U)$$

All three species A, B, and U should be at pH equilibrium, i.e., at the steady state.

D. Resolving Power and Resolution

In isoelectric focusing the resolving power has been defined by Vesterberg and Svensson (1966) to be

$$\Delta pI = 3[d(\text{pH})/dx]\sigma$$

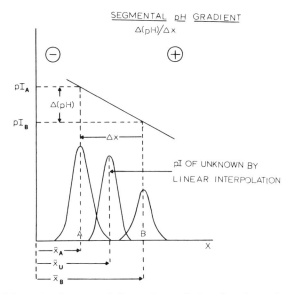

FIGURE 30. Schematic diagram of linear interpolation for the estimation of the segmental pH gradient and the apparent isoelectric point.

Since $\Delta(\text{pH})/\Delta x$ and σ can be obtained digitally in TRANS-IF, the resolving power can be estimated directly. The resolution between two focused peaks can be determined in the same manner as described above for TRANS-MZE experiments.

E. Retardation Coefficient

The apparent diffusion coefficient in polyacrylamide gels is related to the gel concentration by

$$\log D = \log D_0 - C_R T$$

where D_0 is the free diffusion coefficient, T is the gel concentration, D is the apparent diffusion coefficient at any gel concentration T, and C_R is the retardation coefficient obtained from diffusion data. C_R can be measured during the defocusing stage of TRANS-IF experiments (see below) from the slope of the plot $\log D$ vs. T. Thus, TRANS-IF in polyacrylamide gels can provide a measure of molecular size. It is therefore possible that the effective molecular radius $\bar{R}$ and the molecular weight could be estimated by the present method from plots of C_R vs. $\bar{R}$ or vs. mol. wt. in analogy to the Rodbard and Chrambach (1971) plots.

F. Kinetics of Defocusing and Refocusing

Weiss, Catsimpoolas, and Rodbard (1974) presented a restricted theory of the kinetics of TRANS-IF. The principal assumption is that the mobility of the protein is a linear function of position at all times. The TRANS-IF method is assumed to consist of three stages:

1. Focusing, in which the system is allowed to approach the steady-state distribution, for a time t_1.
2. Defocusing, in which the electrical field is abolished for a time t_2. This is assumed to be a pure diffusion process.
3. Refocusing, for a time t_3, in which the field is reapplied, and the distribution again approaches the steady state.

The following assumptions have been adopted to permit a first theoretical approximation:

1. A linear pH gradient is established prior to application of the sample protein. Alternatively, we may assume that the pH gradient is formed very rapidly compared with the kinetics of focusing of the macromolecule of interest.

2. The pH mobility curve of the protein is assumed to be linear. This assumption is valid only for a limited region near the isoelectric point. On the basis of these two assumptions, $p = dM/dx$ is constant. This single assumption could be used in lieu of the above.

3. The electrical field strength ($\mathscr{E}$) is assumed to be uniform throughout the entire separation path. (In lieu of assumptions 1–3, we could simply assume that $pE = dv/dx$ is constant, where v is velocity.)

4. Diffusion and mobility coefficients are assumed to be independent of concentration.

5. Diffusion coefficients are assumed to be independent of pH (at least in the region near the isoelectric point).

6. It is assumed that there are no physical–chemical interactions between the protein and other chemical species present (e.g., ampholytes), and no self-association or protein–protein interactions).

7. Band spreading is governed only by diffusion or by a diffusionlike process. Thus, electrostatic effects are ignored and it is assumed that the protein is perfectly homogeneous with respect to pI, charge, mobility, radius, and diffusion coefficient.

8. No perturbing phenomena such as electroendosmosis, convective disturbances, or precipitation at the isoelectric point are present.

9. If a gel or density gradient is used as a supporting medium, their effects on diffusion coefficients and on mobility are negligible (or, at least constant throughout the gel) and there is no effect on the uniformity of the

electrical field. Thus, the effect of the viscosity gradient which is super-imposed on the density gradient in sucrose-gradient columns is ignored. Similarly, the molecular sieving effects which are present when polyacrylamide gels are used as a supportive medium are ignored.

10. The effect of the boundary condition that there can be no flux of the species of interest through the ends of the gel column, or that there is an abrupt discontinuity of pH at the ends of the column, is ignored. These effects should become insignificant shortly after the start of the experiment.

The above assumptions may be relaxed later to provide a more general-ized and practical theory. For experimental purposes, it is sufficient and convenient to find $\mu_1(\tau)$ and $\sigma^2(\tau)$, i.e., the mean and square of the standard deviation of peak width. These equations are:

1. Focusing: $0 \leq \tau \leq \tau_1$

$$\mu_1(\tau) = \mu_1(0) \exp(-\tau) + Y_0[1 - \exp(-\tau)]$$
$$\sigma_2(\tau) = \sigma^2(0) \exp(-2\tau) + \alpha[1 - \exp(-2\tau)]$$

2. Defocusing: $\tau_1 \leq \tau \leq \tau_1 + \tau_2$

$$\mu_1(\tau) = \mu_1(0) \exp(-\tau_1) + Y_0[1 - \exp(-\tau_1)] = \text{constant}$$
$$\sigma^2(\tau) = \sigma^2(0) \exp(-2\tau_1) + \alpha[1 - \exp(-2\tau_1)] + 2\alpha(\tau - \tau_1)$$

3. Refocusing: $\tau_1 + \tau_2 \leq \tau$

$$\mu_1(\tau) = \{\mu_1(0) \exp(-\tau_1) + Y_0[1 - \exp(-\tau_1)]\} \exp[-(\tau - \tau_1 - \tau_2)]$$
$$+ Y_0\{1 - \exp[-(\tau - \tau_1 - \tau_2)]\}$$
$$\sigma_2(\tau) = \{\sigma^2(0) \exp(-2\tau_1) + \alpha[1 - \exp(-2\tau_1)] + 2\alpha\tau_2\}$$
$$\times \exp[-2(\tau - \tau_1 - \tau_2)] + \alpha\{1 - \exp[-2(\tau - \tau_1 - \tau_2)]\}$$

where L is the column length, x_0 is the position of the isoelectric point, $\alpha = D/(L^2 pE)$, $\tau = pEt$, and $Y_0 = x_0/L$.

A computer-simulation study derived from theory of the time course of the centroid (μ) and σ^2 in TRANS-IF is shown in Figure 31. The centroid approaches the isoelectric point by an exponential decay during focusing and refocusing. With ideal initial pulse loading, and bandwidth (σ^2) increases during focusing, asymptotically approaching the steady-state value.

The parameters that can be determined from these stages of the experi-ment are the apparent diffusion coefficient D and pE (Figure 32). If E and $d(\text{pH})/dx$ are known, the electrofocusing coefficient $dM/d(\text{pH})$ can be estimated. D can be determined from defocusing data and pE from refocusing. In practice, the peak variance (σ^2) is measured as a function of time (Figure

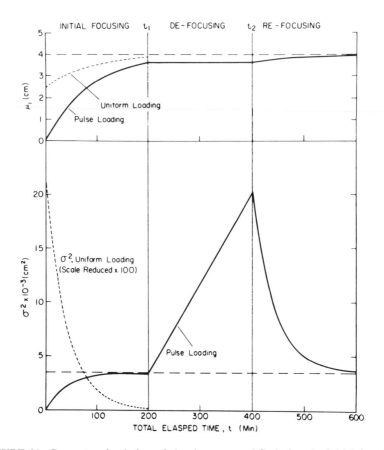

FIGURE 31. Computer simulation of the time course (σ^2) during the initial focusing defocusing, and refocusing stages of TRANS-IF as derived by theory.

33). The apparent diffusion coefficient D is calculated from the defocusing stage as

$$D = \frac{1}{2}\frac{d\sigma^2}{dt}$$

Data from the refocusing stage of the experiment can be analyzed by un-weighted, nonlinear, least-squares curve-fitting methods utilizing the Gauss Newton logarithm for the equation

$$\sigma^2 = (\sigma_0^2 - \sigma_\infty^2) \exp{(-2pET)} + \sigma_\infty^2$$

which provides estimates of pE, $\sigma_\infty^2 = D/pE$, and D, where σ^2, σ_∞^2, σ_0^2 denote peak variance during refocusing, refocusing steady-state, and zero

ISOELECTRIC   DEFOCUSING   ISOELECTRIC   REFOCUSING

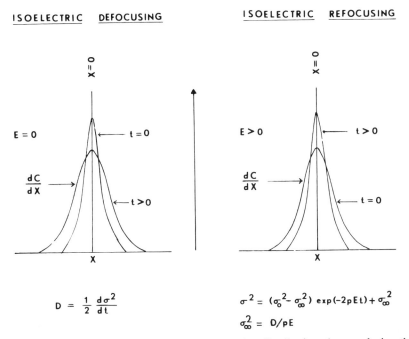

$$D = \frac{1}{2}\frac{d\sigma^2}{dt}$$

$$\sigma^2 = (\sigma_0^2 - \sigma_\infty^2)\exp(-2pEt) + \sigma_\infty^2$$

$$\sigma_\infty^2 = D/pE$$

FIGURE 32. Schematic diagram of the concentration distribution changes during the defocusing and refocusing stages of TRANS-IF.

focusing time (end of defocusing), respectively. Alternatively, pE can be estimated from

$$1n\,(\sigma^2 - \sigma_\infty^2/\sigma_0^2) = -2pET$$

The refocusing steady-state variance (σ_∞^2) is estimated from a graph of σ^2 vs. time as the statistical mean of all the points appearing to be at the steady state. Subsequently, the parameter $\log_{10}(\sigma^2 - \sigma_\infty^2/\sigma_0^2)$ is graphed against $2t$, where t is the duration of refocusing. Least-squares linear regression is utilized to calculate the slope, which is an estimate of $-pE/2.303$ (Catsimpoolas, 1974).

G. Nonideal Effects in TRANS-IF

Although methods have been presented above for the experimental measurement of methodological parameters and physical constants in electrophoresis and isoelectric focusing, these may not represent absolute values. This is especially true for the diffusion coefficient D and the parameter

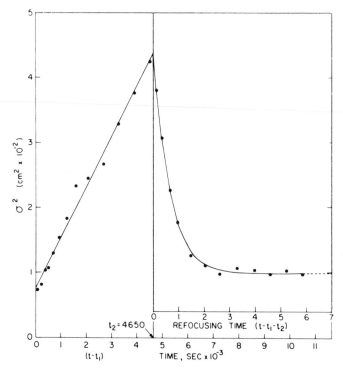

FIGURE 33. Experimental plot of corrected second moment (σ^2) for histidyl–tyrosine vs. elapsed time during defocusing, refocusing and the apparent steady state.

$dM/d(\text{pH})$ (Catsimpoolas, 1973d; Catsimpoolas *et al.*, 1974b). The determination of true values comparable to those obtained by classical methods (e.g., ultracentrifugation and free-boundary electrophoresis) will require corrections in respect to zone load, viscosity, temperature, carrier ampholyte concentration, and polyacrylamide gel concentration. The most serious problem in the interpretation of results by TRANS-IF is the possible uneven distribution of the concentration and conductance of the carrier ampholytes in the isoelectric focusing column. This presents serious problems in measuring the electric field $\mathscr{E}$ and in applying ampholyte-concentration corrections which generally assume uniform distribution of conductance and concentration, respectively. The synthesis of new carrier ampholytes exhibiting stable, linear pH gradients of uniform conductance, concentration, and viscosity may provide valuable impetus to the further development of the TRANS-IF method.

VI. FUTURE DEVELOPMENTS

One area where significant new developments can be expected is instrumentation. The present system which utilizes only absorbance (or percent transmittance) measurements will be expanded in the near future to incorporate fluorescence, variable-angle scattering, and two-wavelength ratio absorbance or fluorescence. The sensitivity of detection can be improved by the use of a chopper and lock-in amplifiers. Alternatively, photon-counting techniques can be considered. Since the linear movement of the column can be stopped at any peak of interest, a second dimension scanning, e.g., at variable wavelengths or angle of scattering, is feasible. In this respect, the use of optical multichannel analyzers (O.M.A., Princeton Applied Research) will be of particular interest. The desirable feature of multiple-column analysis cannot be overlooked, especially in experiments where a number of variables have to be examined simultaneously (e.g., variable $\%T$ for the derivation of K_R, zone-load correction for diffusion experiments, etc). Again an automatic $x-y$ mechanical scanner has been constructed and is currently evaluated in this laboratory. Eventually, an on-line dedicated computer will be a necessary addition to the system for data acquisition and processing, and for the automatic control of the instrument.

Further general advances in testing old and new electrophoretic theories and systems by TRANS electrophoresis can be safely predicted. Additionally, the method can be easily adapted to evaluate interaction and binding phenomena and possibly be expanded to include immunodiffusion reactions in gels. Finally, the TRANS instrument will be a logical candidate for the computerized electrophoretic analysis of biomolecules and bioparticles in the clinical laboratory.

ACKNOWLEDGMENTS

Part of the above work was supported by a contract from the National Cancer Institute (NO1-CB-43928). The author also wishes to acknowledge the valuable contributions of Dr. A. Chrambach, Dr. A. L. Griffith, Dr. D. Rodbard, Dr. W. W. Yotis, Dr. J. Wang, Dr. G. H. Weiss, Mr. B. E. Campbell, Mrs. J. Kenney, Mr. K. Lohse, and Mr. J. Williams in the development of the TRANS electrophoresis method.

VII. REFERENCES

Borris, D. P. and Aronson, J. N. (1969) *Anal. Biochem.* **32**: 273.
Brakke, M. K., Allington, R. W., and Langille, F. A. (1968) *Anal. Biochem.* **25**: 30.
Brumbaugh, E. E. and Ackers, G. K. (1968) *J. Biol. Chem.* **243**: 6315.
Catsimpoolas, N. (1971*a*) *Sep. Sci.* **6**: 435.
Catsimpoolas, N. (1971*b*) *Anal. Biochem.* **44**: 411.
Catsimpoolas, N. (1971*c*) *Anal. Biochem.* **44**: 427.
Catsimpoolas, N. (1973*a*) *Ann. N.Y. Acad. Sci.* **209**: 65.
Catsimpoolas, N. (1973*b*) *Fed. Proc.* **32**: 625.
Catsimpoolas, N. (1973*c*) *Anal. Biochem.* **54**: 66.
Catsimpoolas, N. (1973*d*) *Anal. Biochem.* **54**: 79.
Catsimpoolas, N. (1973*e*) *Anal. Biochem.* **54**: 88.
Catsimpoolas, N., ed. (1973*f*) *Ann. N.Y. Acad. Sci.* **209**: 1–529.
Catsimpoolas, N. (1974) in *Electrophoresis and Isoelectric Focusing in Polyacrylamide Gel* (R. C. Allen and H. R. Maurer, eds.) p. 174, Walter de Gruyter, Berlin.
Catsimpoolas, N. (1975*a*) in *Isoelectric Focusing* (J. P. Arbuthnott and J. A. Beeley, eds.) p. 58. Butterworths, London.
Catsimpoolas, N. (1975*b*) in *New Developments in Separation Methods* (E. Grushka, ed.) Marcel Dekker, New York.
Catsimpoolas, N. and Campbell, B. E. (1972) *Anal. Biochem.* **46**: 674.
Catsimpoolas, N. and Griffith, A. L. (1973) *Anal. Biochem.* **56**: 100.
Catsimpoolas, N. and Wang, J. (1971*a*) *Anal. Biochem.* **39**: 141.
Catsimpoolas, N. and Wang, J. (1971*b*) *Anal. Biochem.* **44**: 436.
Catsimpoolas, N., Campbell, B. E., and Griffith, A. L. (1974*a*) *Biochim. Biophys. Acta* **351**: 196.
Catsimpoolas, N., Yotis, W. W., Griffith, A. L., and Rodbard, D. (1974*b*) *Arch. Biochem. Biophys.* **163**: 113.
Catsimpoolas, N., Griffith, A. L., Williams, J. M., Chrambach, A., and Rodbard, D. (submitted).
Chrambach, A. and Rodbard, D. (1971) *Science* **172**: 440.
Dravid, A. R., Fredén, H., and Larsson, S. (1969) *J. Chromatogr.* **41**: 53.
Easton, D. M., Lipner, H., Hines, J., and Leif, R. C. (1971) *Anal. Biochem.* **39**: 478.
Fawcett, J. S. (1970) *Prot. Biol. Fluids Proc. Colloq.* **17**: 409.
Hanlon, S., Lamers, K., Lauterbach, G., Johnson, R., and Schachman, H. K. (1962) *Arch. Biochem. Biophys.* **99**: 157.
Hjertén, S. (1967) *Free Zone Electrophoresis*, Almquist and Wiksells, Uppsala.
Hochstrasser, H., Lerner, H., and Skeggs, L. T., Jr. (1967) in *Electrophoresis* (M. Bier, ed.) Vol. 2, p. 473, Academic Press, New York.
Johnsson, M., Pettersson, S., and Rilbe, H. (1973) *Anal. Biochem.* **51**: 557.
Jovin, T. M. (1973) *Ann. N.Y. Acad. Sci.* **209**: 477.
Jovin, T. M., Dante, M. L., and Chrambach, A. (1970) *Multiphasic Buffer Systems Output*, National Technical Information Service, Springfield, Va., PB Nos. 196085–196092 and 203016.
Loening, U. E. (1967) *Biochem. J.* **102**: 251.
Olivera, B. M., Baine, P., and Davidson, N. (1964) *Biopolymers* **2**: 245.
Ornstein, L. (1964) *Ann. N.Y. Acad. Sci.* **121**: 321.
Radola, B. J. and Delincée, H. (1971) *J. Chromatogr.* **61**: 365.

Ressler, N. (1967) in *Electrophoresis* (M. Bier, ed.) Vol. 2, p. 493, Academic Press, New York.

Rodbard, D. and Chrambach, A. (1971) *Anal. Biochem.* **40**: 95.

Rodbard, D. and Chrambach, A. (1974) in *Electrophoresis and Isoelectric Focusing in Polyacrylamide Gel* (R. C. Allen and H. R. Maurer, eds.) p. 28, Walter de Gruyter Berlin.

Rodbard, D. Chrambach, A., and Weiss, G. H. (1974) in *Electrophoresis and Isoelectric Focusing in Polyacrylamide Gel* (R. C. Allen, and H. R. Maurer, eds.) p. 62, Walter de Gruyter, Berlin.

Svensson, H. (1961) *Acta Chem. Scand.* **15**: 325–341.

Vesterberg, O. and Svensson, H. (1966) *Acta Chem. Scand.* **20**: 820.

Watkin, J. E. and Miller, R. A. (1970) *Anla. Biochem.* **34**: 424.

Weiss, G. H., Catsimpoolas, N., and Rodbard, D. (1974) *Arch. Biochem. Biophys.* **163**: 106.

IMMUNODIFFUSION   3

ALFRED J. CROWLE

I. INTRODUCTION

Immunodiffusion is an analytic technique in which reactants diffuse to intermingle with each other and react immunologically. Generally, the primary reagent will be antiserum and the substance to be analyzed will be antigen; but sometimes these roles are reversed, antigen being used to study antibodies. Immunodiffusion tests are performed in semisolid media, usually gels, to allow reactants to diffuse while preventing convective mixing, and to support accumulation of the antigen–antibody reaction product, usually a precipitate, for observation.

There are many varieties of immunodiffusion, these differing from each other and being named by whether one or both antigen and antibodies move during the test, by how these reactants move, and by what may be done to these reactants before they come in contact with each other (Crowle, 1973). For instance, in single-diffusion tests one reactant moves while the other remains sessile, in an electroimmunodiffusion test reactants are mixed electrophoretically, in two-dimensional electroimmunodiffusion, one of the reactants is fractionated electrophoretically in one plane before it is electrophoresed in another to react with the other.

II. ANTIGEN–ANTIBODY REACTIONS

Antigen–antibody reaction is an essential part of all immunodiffusion tests, and, with only a few exceptions, is a precipitation reaction. The follow-

ALFRED J. CROWLE, Webb-Waring Lung Institute and Department of Microbiology, University of Colorado School of Medicine, Denver, Colorado 80220.

ing is a brief review of the natures of antigens and antibodies, of antisera, and of the antigen–antibody precipitation reaction.

Antigens are molecules which will stimulate an animal injected with them to make antibodies to them. They are macromolecules, usually proteins, and the more foreign they are to the animal the more antigenic they will be. Other kinds of macromolecule like polysaccharides, nucleic acids, and sometimes lipids also can be antigenic. Smaller molecules of various kinds can be made to stimulate antibody formation against themselves, but they are called haptens rather than antigens because in order for them to elicit this antibody formation they must be complexed with an antigenic carrier macromolecule.

The antibodies which an animal can make when stimulated with antigen are of many kinds and characteristics. Those used in immunodiffusion tests are found among the gamma globulins of the body fluids, and of these fluids blood serum is the most frequently employed. There are exceptions such as in the mouse in which more acites fluid may be available than serum (Krøll, 1970). From the immunized animal one obtains an antiserum which, in turn, contains gamma-globulin antibodies that can precipitate the same kind of antigen as was used to induce antibody formation but which does not precipitate unrelated antigens. This high specificity of the primary reagent for immunodiffusion tests, antiserum, and that the reagent can be tailor-made for this specific reaction are unique advantages of immunodiffusion tests over other kinds of analyses for complex macromolecules such as proteins.

Antibodies and antiserum are not equivalent either in fact or in use. The word "antibodies" refers to antigen-reactive gamma globulins, or immunoglobulins, whereas "antiserum" refers to the whole serum containing such antibodies. Some antibodies may be able to precipitate antigen and therefore are known as precipitins; others may not. Still others can react with antigen in such a way as to interfere with interaction between the antigen and precipitins, and therefore with precipitation. Hence, antiserum reacts with antigen to produce a net effect of all of its antibodies specific for the antigen. In addition to this, serum contains substances which are not antibodies themselves but which can participate in poorly understood ways in antigen–antibody reactions, usually enhancing them, but sometimes suppressing them (Crowle, 1973, Chapter 1). Either antiserum or antibodies purified from antiserum can be used in immunodiffusion tests, but their respective results are likely to differ somewhat because they are not equivalent. For most tests antiserum seems to be the better choice.

The precipitins usually reacting in immunodiffusion tests are divalent. Each combining site on one precipitin molecule has the same specificity for antigen as the other combining site. Most antigens are multivalent, generally having many different determinants which can be recognized by matching antibody molecules. When antigen and its antibodies are mixed antibodies collide with antigenic sites and complex firmly with those they recognize in what is called the primary reaction. By means of its two combining sites an antibody molecule can connect two molecules of the same antigen. Multiple successive connections of this sort form antigen molecules and their antibodies into a latticework, constituting the secondary reaction. The ultimate complex continues to enlarge, eventually becoming insoluble and precipitating. These antigen–antibody complexes cannot grow large enough to precipitate unless antibodies which recognize several different reactive sites on the antigen are present to cooperate with each other, or unless the antigen has repeating units of identical determinant, as dipeptide proteins would and polysaccharides might. Obviously, then, high titers of divalent antibodies are not sufficient to enable an antiserum to precipitate its antigen; the antiserum must contain antibodies with an adequate variety of antigenic specificities.

When several antigens are mixed in liquid with antiserum containing antibodies to more than one of the antigens, the resulting precipitate is a mixture of lattices of all the different reacting systems. There is no indication in such a test of how many antigens were in the original mixture. But, if the antiserum is gelled in a test tube and overlaid with a solution containing several antigens, the result will be different. Antigens and antibodies can no longer mix freely; they will encounter each other only as the antigens diffuse into the antiserum-charged gel, as shown in Figure 1. Here each different

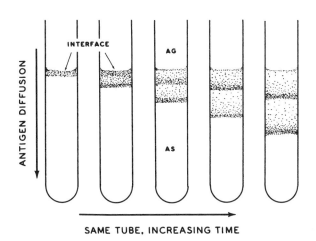

SAME TUBE, INCREASING TIME

FIGURE 1. Diagrammatic representation of an Oudin-type single-diffusion tube test in which antigens diffuse from a solution into antiserum-charged gel where they form independent waves of precipitate which descend progressively farther into the antiserum gel with increasing time.

antigen will be precipitated apart from any other by just those antibodies specific for it. One reason for this independent precipitation has already been mentioned, namely the high specificity of antigen–antibody reactions. One precipitating system will not interfere with nor participate in precipitation of another. A second reason will be evident from the physical nature of this type of test: It is that most probably no two populations of antigen will be diffusing at equal rates into the antiserum-charged gel because they probably would not be present in the mixture in identical concentrations and they would be unlikely to have identical diffusion coefficients. A third reason requires additional explanation of antigen–antibody precipitation.

Antigen and antibodies can react in many different ratios. When they are mixed with antigen in excess, they react, but precipitation is poor or absent because few antibodies are available to bridge between antigen molecules

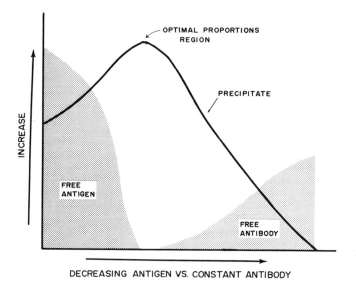

DECREASING ANTIGEN VS. CONSTANT ANTIBODY

FIGURE 2. Graphic depiction of the relationships between antigen, antibodies, and antigen–antibody precipitate in a tube precipitation test using liquid medium. As decreasing quantities of antigen are mixed with constant quantities of antibody, there are shifts from conditions where free antigen is found in the supernatant fluid above moderate amounts of precipitate, through the optimal-proportions region where precipitate is maximum and neither antigen nor antibody can be found in the supernatant fluid, to progressively decreasing quantitites of precipitate in which, from lack of antigen, one begins eventually to find some free antibody in the supernatant fluid. The figure is meant especially to show that antigen–antibody complexing develops in several different ratios, that for each antigen–antibody system there is an optimal ratio at which precipitation is best, and that antigen–antibody precipitation can be impeded or reversed by excesses of either reactant.

to form a lattice (Figure 2). Similarly, although to a lesser extent, when antibodies are excessive, precipitation is poor. At an optimal ratio of antigen to its antibodies, precipitation will be strongest and fastest, and this optimal ratio depends on several characteristics of both the antibodies and the antigen. Consequently, two different antigen–antibody systems seldom have identical optimal proportions, and in the single-diffusion tube test shown in Figure 1 this difference for each of the various precipitating systems contributes to the effects of the other differences to make each antigen–antibody system precipitate in a different plane of the antiserum-charged gel.

On observing this single-diffusion tube test for several hours, one will notice, as indicated in Figure 1, that the forming bands of precipitate are not static; they appear to move down into the gel. This observation represents another important characteristic of antigen–antibody precipitation which affects immunodiffusion tests, that the precipitation is reversible, especially in the presence of excess antigen. The band of precipitate appearing to move through the antiserum-charged gel is not itself moving; it is being dissolved from behind and is precipitating anew at its advancing front, somewhat analogous to an ocean wave breaking on a beach.

III. IMMUNODIFFUSION

That antigen–antibody precipitation can develop in a gelled aqueous medium has been known for nearly 70 years (Crowle, 1973, Chapter 8). But use of this knowledge did not begin in earnest until 1946 because the multiple bands of precipitate which frequently are developed in an immuno-diffusion test by independent antigen–antibody systems generally were misinterpreted as multiple precipitates produced by just one precipitating system. In 1946 Oudin began publishing evidence that each band in a single-diffusion tube test was formed by a separate antigen–antibody system, and in 1948 Ouchterlony confirmed and extended these observations and greatly facilitated techniques for studying and comparing these different systems by inventing the radial double-diffusion test, also frequently called the Ouchterlony test.

In Oudin's single-diffusion tube test antigen solution is layered in a small tube over agar gel containing a relatively weak concentration of antiserum. Disks or bands of precipitate initially form at the reactant interface, but as antigen diffuses rapidly into the underlying gel these bands of precipitate appear to migrate through the antibody gel as already shown in Figure 1. In this test, then, only one reactant diffuses significantly, and this diffusion is linear.

In Ouchterlony's double-diffusion plate test, agar gel is cast on a flat surface, wells are cut in it, and the wells are filled with antigen and antiserum. Each reactant diffuses radially from its source in an ever-widening invisible disk until somewhere between the two respective sources the edge of the disk of antibodies meets the oppositely moving disk of antigen molecules, and antigen and antibodies begin to react, as depicted in Figure 3. A thin line of precipitate develops where the two reactants attain optimal proportions near where they first met and grows laterally as larger proportions of the still expanding disks of antigen and antibodies merge into each other. Since there is excess antigen on the antigen-source side of this developing band of precipitate and excess antibody on the antiserum-source side, the precipitate itself represents a dynamic barrier to penetration beyond it of either reactant involved in its formation. On the other hand, it is not a physical barrier, for unrelated antigens and antibodies can diffuse through it nearly as easily as they diffuse through the agar gel itself. Therefore, several bands of precipitate can form in separate planes of the gel between antigen and antiserum sources if the antigen solution being tested contains several different populations of antigen and the antiserum well was charged with an antiserum containing the correspondingly reactive populations of antibodies. This is illustrated in Figure 4.

This dynamic impermeability of a precipitin band to antigen and antibodies participating in the band's formation and permeability to unrelated antigens and antibodies is one of the most basic and important attributes of immunodiffusion tests. It explains reactions of identity, partial identity, and nonidentity not only in comparative Ouchterlony tests, but also in various kinds of single-diffusion tests, in immunoelectrophoresis, and in electroimmunodiffusion. The reaction of identity develops in an Ouchterlony plate

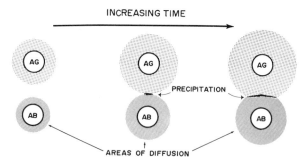

FIGURE 3. This figure shows that development of antigen–antibody precipitate bands in a double-diffusion plate (Ouchterlony) test depends on expanding radial diffusion of antigen and antibodies, and their overlap, mixture, and reaction in some plane between their two sources.

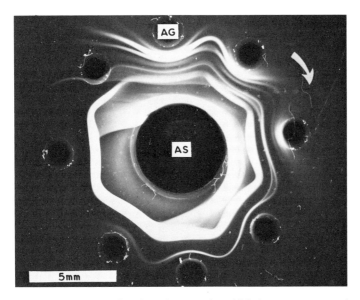

FIGURE 4. A micro-double-diffusion plate test in which human serum antigen was reacted with horse antiserum to human serum. This test was both qualitative and quantitative. A 10% concentration of antigen charged the 12-o'clock well, and each well in a clockwise direction was charged with the same volume but half the antigen concentration of the preceding well (i.e., 5%, 2.5%, 1.25%, etc.). Some interpretations of this test are given in the text. Note that the central well in this microtest is about 5 mm in diameter.

when two solutions of identical antigen are allowed to diffuse from juxtaposed wells toward a third well charged with antiserum (Figure 5). A band of precipitate forms between each of the two antigen sources and the common antiserum origin, and when these have grown laterally far enough toward each other to meet, they grow no further but fuse in the reaction of identity. By contrast, if different antigens are compared with each other in this manner, when their respective lines of precipitate meet, they continue to grow across and through each other in the reaction of nonidentity. Sometimes antigens which are being compared are related but not identical; for instance, digested fragments of an antigen could be compared with the native antigen. This results in a reaction of partial identity in which the line of precipitate formed by the native antigen continues to grow past the point where it intersects the line formed by the digested antigen, but the line formed by the latter stops growing at that point. The reason is that the antiserum has antibodies to determinants found on both types of antigen and also to additional determinants found on the native antigen but missing from the digested antigen. These latter antibodies are able to diffuse through the precipitate being formed

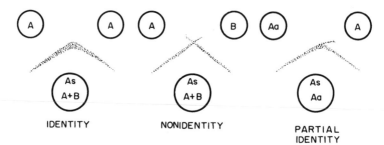

FIGURE 5. Diagram showing the characteristics in comparative double-diffusion plate tests of reactions of serological identity, nonidentity, and partial identity when two solutions of antigen are compared by simultaneous reaction with an antiserum containing antibodies to hypothetical antigens A and B. As explained in the text, each antigen must be multivalent and the antiserum must contain antibodies to each of several antigenic determinants for these reactions to develop as depicted.

by the incomplete antigen and to react with the complete antigen on the other side of the precipitate forming what is called a spur.

Simple as are the explanations of these three kinds of reactions, one must always accept their meaning with considerable caution. For example, the reaction of identity does not really mean that compared antigens are identical; it means only that the antiserum being used for their comparison cannot distinguish between them. Obviously, then, an immunodiffusion test is only as good as the antiserum that is used in it; and since antisera produced against any antigen by different animals or by different regimens have different properties, any result obtained with an immunodiffusion test must be reported with due mention of the kind of antiserum that was used.

In interpreting immunodiffusion patterns one must always keep in mind the characteristics of antigens and antibodies, and of their interactions. For instance, a reaction of identity for two compared antigens if not necessarily meaning that they are identical at least means that they are closely related. But a reaction of identity for two compared antisera with respect to a single antigen does not at all mean that the antisera contain identical or even closely related antibodies. The reason for this puzzling distinction in interpretation is that antibodies are divalent and antigen is multivalent. The antibodies in one of the compared antisera may be specific for several determinants on the antigen, while the antibodies in the other antiserum can react with an entirely different set of determinants. Yet the two antisera develop a reaction of identity because they are reacting with a single, common population of antigen molecules. Hence, a reaction of identity when antisera are being compared means that both antisera can react with the same antigen, but it neither indicates that the antibodies in the antisera are identical nor how closely related the antigenic specificities of the compared antisera are. This is a common and probably frequently overlooked problem in antiserum standardizations.

Thus, although reactions of identity, partial identity, and nonidentity are indispensable indicators for and inseparable characteristics of all kinds of immunodiffusion test, they can and frequently do develop falsely and therefore must be interpreted circumspectly.

I have discussed the nature of antigen–antibody precipitation in aqueous solutions and in Oudin single-diffusion tubes and Ouchterlony double-diffusion plates mostly from the point of view of the primary reactants. Let me consider next some factors which affect choices of reactants to be used and methods for using them. One must select an appropriate antiserum (cf. Crowle, 1973, Chapter 2). If the purpose is to detect as many antigens as possible in a mixture of antigens, such as in characterizing studies or in attempts to define purity, then the antiserum must be polyvalent and strong. The utility of such an antiserum is well illustrated in Figure 4. It should be obtained from a hyperimmunized animal, i.e., an animal which has been exposed repeatedly to the antigen mixture being analyzed and is making precipitins vigorously to all the antigens in the mixture.

By contrast, if the purpose of a test is to detect one antigen among several which may be closely related, as in forensic distinction between human and primate blood, the antiserum should be obtained from a weakly immunized animal, preferably one closely related to that from which the antigen was obtained.

Certain kinds of antisera may be better than others for reasons other than how specifically or extensively they react with antigen. Rabbit or goat antisera are better than horse antisera for quantitative experiments because antigen–antibody precipitates formed by horse antisera are unstable. Chicken antisera are more useful for studying antigens that are soluble only in relatively high salt concentrations because precipitins in chicken antisera function well at such salt concentrations (e.g., 1.5 M NaCl). For certain kinds of electro-immunodiffusion tests horse and mouse antisera are better than rabbit anti-sera because their precipitins have gamma-1 electrophoretic mobility whereas rabbit precipitins have gamma-2 mobility.

Immunodiffusion tests must be performed in some kind of anticonvection medium which does not interfere with reactant diffusion. Agar, agarose, gelatin, cellulose acetate, and polyacrylamide have been used (Crowle, 1973; Williams and Chase, 1971). Of these, agar has been the most popular, but currently it is being displaced by agarose because the latter is more uniform, cleaner, and more inert. Agar is readily available and easy to use, requiring only that it be dissolved in a 1% concentration in boiling water and then cooled to room temperature. Because of the agaropectin that it contains, however, it has a strong negative charge which in electrophoresis causes strong electroendosmotic flow of water and antigens dissolved in the water. This strong electronegativity also precludes analyses of positively charged reactants and also staining the gel with basic dyes because these combine with the agaropectin. When agaropectin is removed from agar, agarose remains. This forms a better, clearer gel at slightly lower concentrations than agar, it is nearly electroneutral and compatible with positively charged reactants or dyes, and it is as easy to use as agar.

Gelatin is seldom used for several reasons (Crowle, 1973), two of which are that it interferes with antigen–antibody precipitation and that it cannot be used with conventional immunodiffusion stains. Cellulose acetate offers advantages of reactant economy, anticonvection medium uniformity, and inertness, but one cannot see antigen–antibody precipitation in cellulose acetate until it has been washed and stained, and some antigen–antibody precipitates are unstable enough to disappear during washing and staining procedures.

At first consideration polyacrylamide would seem to be an ideal medium for immunodiffusion tests because it is inert, very stable, and water-clear. Occasionally it is used, but it has never gained the popularity of either agar or agarose because it is harder to prepare and harder to handle, and antigen–antibody precipitation does not seem to develop as well in it (Crowle, 1973). Furthermore, if it is prepared as it is used in disk electrophoresis it interferes with diffusion of antigens and antibodies.

The best medium for most immunodiffusion tests, therefore is agarose. It is used most frequently at 1 or 1.5%.

Many different formulas have been published for the buffer in which agar or agarose is dissolved for an immunodiffusion test (Crowle, 1973; Williams and Chase, 1971). Some general rules are better than specific recommendations for suggesting which are the better choices. Antigen–antibody precipitation usually is best at physiologic ionic strength and pH, so most of the tests not using electrophoresis are set up accordingly. But precipitation will develop adequately up to about pH 9, and nonspecific precipitation does not become objectionable until the pH falls below 6; adequate precipitation also can be obtained in salt concentrations ranging from equivalents of 0.02 N to 1.5 N sodium chloride. Consequently, when variations in buffer composition are necessary to perform certain operations in immunodiffusion tests they generally can be made within these limits and still provide satisfactory results. For example, high salt concentrations may be necessary to maintain solubility of certain kinds of antigen; low salt concentrations must be used in immunoelectrophoresis and most forms of electroimmunodiffusion to prevent overheating of the gel during electrophoresis. A pH of 8.6 is commonly employed in both immunoelectrophoresis and single electroimmunodiffusion to minimize movement toward the cathode of antigens and antibodies.

The most popular buffers for immunodiffusion tests are barbital and phosphate, but there are many others which can be used. Some have known disadvantages. Buffering agents which chelate divalent cations can suppress antigen–antibody precipitation. Some like acetate and borate can complex with antigens. Still others are selectively incompatible. For example, precipitation of alkaline phosphatase by its antibodies will not develop in agarose buffered with phosphate but will in barbital-buffered gel (Crowle and Atkins, 1974).

A very important discovery has been made recently about the composition of immunodiffusion solvents. On the one hand it will require some restructuring of traditional ideas concerning the mechanisms of antigen–antibody precipitation; on the other, it makes immunodiffusion analyses applicable to a vast new area of investigation which has remained relatively untouched with analyses offering the resolution that immunodiffusion tests are capable of. This discovery is that antigen–antibody precipitation can develop very well in concentrations of nonionic detergents like Triton X-10 and Berol EMU-043 that are sufficient to solubilize membrane antigens. Concentrations of 0.5–2.0% have reportedly been used by several laboratories (cf. Bjerrum and Lundahl, 1974). In limited experiments with two-dimensional electroimmunodiffusion (EID) we have found concentrations of Triton X-100 at which precipitation becomes inhibited in different antigen–antibody systems (Crowle and Atkins, 1974). Most readily withstand 1%, many withstand 2%, while some precipitate adequately in 8%. In accompanying experiments we have found that 1 M and 2 M urea also can be used for many antigen–antibody systems, and 4 M can be used for some. Parenthetically, one cannot add any but the lowest concentrations (e.g., <0.5%) of either Triton X-100 or urea to a solution of agarose which is about to be cast because gelling will be impaired or prevented. One must cast the gel first and then charge it with solubilizing agent by diffusion. A few papers attesting to the exciting analyses which have become possible with realization that these solubilizers can be employed in immunodiffusion tests have already been published. For instance, one reports detection by electroimmunodiffusion of 19 different human erythrocyte membrane antigens (Bjerrum and Lundahl, 1974).

This review so far has dwelt on basic information on the nature of immunodiffusion tests and the reactants and reagents used in them. Since its objective is to present ways in which these tests can be used to analyze and separate proteins, it will discuss techniques, starting with a brief consideration of the varieties of immunodiffusion tests to point out their particular advantages and special uses.

Currently, radial single-diffusion plate tests are the most widely used of immunodiffusion techniques for quantitating proteins. Marketing test kits to quantitate human serum antigens has become a profitable business. This kind of test is begun by incorporating antiserum in warm agarose solution and then casting it as a thin, very uniform gel. Then small wells are cut in the gel and each is charged with antigen solution. The plate is set aside to allow time for antigen in each well to diffuse radially out into the antiserum-charged gel to form a disk of precipitate. This expands until the antigen comes to equilibrium with the concentration of antiserum which has been incorporated in the gel. As Figure 6 suggests, at equilibrium, the size of the resulting disk of precipitate will be proportional to the original quantity of antigen placed in an antigen well and inversely proportional to the concentration of antiserum in the surrounding gel. Hence, with appropriate standardization, the radial single-diffusion plate test can be used for quantitating whatever antigen the antiserum used in the gel will detect. However, it cannot readily be used for studying more than one antigen at a time because multiple antigen–antibody systems will form superimposed disks of precipitate, making individual disks difficult to measure accurately and raising uncertainties as to which disk represents which antigen. This limitation in turn means that to use the radial single-diffusion plate test one should have a monospecific antiserum, and frequently this can be difficult to prepare. Advantages to the technique are simplicity, economy, and being able to measure directly the quantity of one single kind of protein among many in a solution.

Radial double-diffusion plate tests, or Ouchterlony tests, are the most familiar of immunodiffusion techniques (cf. Figure 5). Miniaturized versions

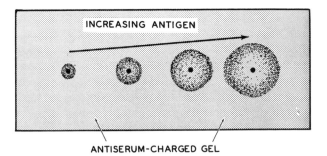

FIGURE 6. Diagram showing some important aspects of the quantitative radial single-diffusion test. The antigen solution to be tested is allowed to diffuse from wells cut in antiserum-charged gel until the antigen comes to serologic equilibrium with its antibodies by forming a stable disk of precipitate whose size is proportional to the quantity of antigen placed in a well and inversely proportional to the concentration of antibodies in the surrounding gel.

have become the most popular because they are economical, faster and more sensitive than large versions, offer better resolution of different antigen–antibody precipitates forming at the same time, and are easier to wash, stain, and preserve. The simplest is set up by casting a 1% agarose gel evenly on a level microscope slide, punching an appropriate pattern of wells with a 16- or 18-gauge hypodermic needle with its tip modified for well cutting, and filling the wells with antigen solutions or antisera. However, a much improved technique offering better sensitivity, resolution, and reproducibility is nearly as easy (Crowle, 1973, p. 291). A thin, flat agarose gel is cast between two microscope slides separated by thin spacers, one slide is drawn off of the gel, and Plexiglas templates with an appropriate pattern of wells already cut in them are slid or laid upon the thin gel. Wells in the templates are filled with reactants, and the slides are placed in a refrigerator for 24 hr for immunoprecipitation to develop.

As Figure 4 shows, micro-double-diffusion template tests are capable of resolving 10–20 different antigen–antibody systems. Since they are more sensitive than other immunodiffusion tests, they are used for such purposes as testing an antigen solution for purity, for monitoring purification at successive stages, for screening antisera for range or strength of reactivity, and for detecting traces of antigen or precipitins. The antigen analyzed by the test shown in Figure 4 was supposed to have been "pure," being crystallized human serum albumin. But numerous contaminants obviously were detected with the hyperimmune horse anti-human-serum antiserum used in that test.

These tests also can be used for quantitating either antigens or precipitins and, unlike radial single-diffusion plate tests, they can quantitate several antigen–antibody systems at once. The test in Figure 4 shows, for instance, that most of the contaminating antigens were present in very low concentration: while readily apparent in a 10% solution of the albumin (12-o'clock well), most have become inapparent in a 0.625% solution (fourth well, clockwise). This observation and other interpretative data suggest that most were present in the original albumin preparation at no more than about 0.004%. Since there were five of these minor antigens and a major one forming a precipitate even at the 9-o'clock well (indicating that this contaminant was present at about 0.6%), the albumin being studied was roughly $100 - (5 \times 0.004\%) - 0.06\% = 99.92\%$ pure.

These and other forms of radial double-diffusion plate test are also used for critical comparisons of antigenic specificity, as in demonstrating reactions of identity, partial identity, and nonidentity discussed earlier.

Although one can see individual antigen–antibody systems precipitating separately from each other in double-diffusion plate tests, one cannot readily identify them. For this purpose, and for increased resolution, compound immunodiffusion tests like immunoelectrophoresis and electro-immunodiffusion should be used and will be discussed below.

One characteristic which may help identify a protein antigen is its diffusion coefficient, and this can be estimated by a double-diffusion plate test in which linear fronts of antigen and antibodies diffuse at right angles across each other. In its simplest form this test is set up by laying antigen- and antiserum-saturated strips of filter paper in L or T configurations on a thin layer of agarose gel. The two strips should not touch each other where they converge. If the antigen has the same diffusion coefficient as its anti-bodies, and the two reactants are used in near optimal proportions, the line of precipitate which it forms with its antibodies will be at 45° between the two reactant sources. If the diffusion coefficient of antigen is greater than that of its antibodies, the line of precipitation will form at an angle more acute toward the antiserum source; if it is less, this angle will be more acute toward the antigen source. Since the diffusion coefficient of precipitins for most commonly used antisera is well known, this physical characteristic can be calculated for any antigen from its angle of precipitation in this type of immunodiffusion test. Recently, this technique has been adapted to use with immunoelectrophoresis for simultaneous analysis of several major serum antigens as is illustrated in Figure 7 (Afonso et al., 1972). Diffusion coefficients estimated by this technique agree closely with those determined by more traditional techniques.

IV. IMMUNODIFFUSION COMBINED WITH ELECTROPHORESIS

Had immunodiffusion technology never developed beyond the double-diffusion plate test, immunodiffusion might not have become the primary tool for protein analyses that it is today. Fortunately, in 1953, only seven years after Oudin had demonstrated the true meaning of multiple antigen–antibody precipitation in single-diffusion tubes and five years after Ouchter-lony had described his double-diffusion plate test, Grabar and Williams (1953) greatly expanded the applicability and usefulness of immunodiffusion techniques by inventing immunoelectrophoresis. During the two decades which followed immunoelectrophoresis was so intensely exploited that it overshadowed another basic improvement in immunodiffusion technology which is only now about to be equally exploited, namely electroimmuno-diffusion (EID) in which mixing of reactants is electrophoretic instead of by diffusion. The first double-EID test, in which both reactants are electro-phoresed within a gel, was described by Crowle in 1956; Ressler described the first gel single-EID test in 1960.

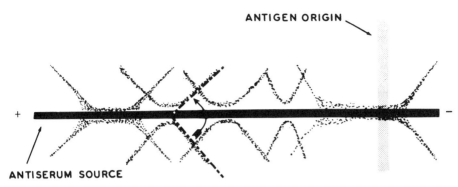

FIGURE 7. Diagrammatic illustration of how diffusion coefficients of several different antigens can be estimated from a single orthogonal immunoelectrophoresis test. If the diffusion coefficient of antibodies being used is known, then the diffusion coefficient of antigen being precipitated can be estimated from the angle at which its precipitate forms between the horizontal antiserum source and the vertical location of the antigen at the end of its electrophoretic migration. Diagram prepared from photographic information published by Afonso *et al.* (1972).

Essentially, immunoelectrophoresis is a double-diffusion plate test with much improved capacity for resolving individual antigens in such complex mixtures as human serum, and, at the same time, for characterizing them electrophoretically (cf. Figure 7). This is because the antigens are electrophoresed before they are allowed to react by diffusion with their antibodies in a polyvalent antiserum. Thus, in addition to detecting such antigens, it indicates whether they are, for instance, alpha, beta, or gamma globulins; and because antigens are separated from each other electrophoretically before they mix and precipitate independently by diffusion, they are much better resolved from each other than in ordinary double-diffusion tests. Immunoelectrophoresis is more difficult than single- or double-diffusion tests, but this disadvantage is easily offset for qualitative analyses by its greater versatility.

The reader probably is less acquainted with electroimmunodiffusion tests than with immunoelectrophoresis because papers describing their use are still relatively rare. However, he should learn about them because they are as much of an analytic improvement over immunoelectrophoresis as immunoelectrophoresis was over the double-diffusion test. One-dimensional single-EID tests can quantitate a protein more accurately than a single-diffusion plate test, and can do so in a few hours instead of a few days. Two-dimensional single EID increases resolution of complex mixtures of antigens about threefold over that attainable by immunoelectrophoresis; and at the same time it quantitates the antigens it detects and reveals electrophoretic variations in them that cannot be seen in immunoelectrophoresis. One-dimensional double EID is more sensitive than double-diffusion plate tests and can provide results within a few minutes instead of hours or days.

In single EID, one reactant, usually antiserum, is incoporated into agarose gel and the other is then electrophoresed into this gel, where it forms a lancet-shaped precipitate as shown in Figure 8. The reaction can be complete within as little as one hour. Quantitative information derives from the fact that if the lancet of precipitate comes to a sharp point its length is directly proportional to the concentration of antigen and inversely proportional to the concentration of antibodies precipitating it. Sometimes this test is called the rocket test because of its characteristic shape of antigen–antibody precipitation.

If several antigens are present in the antigen solution being analyzed by one-dimensional single EID, several lancets or loops of precipitate will form confusingly within or upon each other. By separating the antigens from each other laterally by electrophoresis before they are electrophoresed into

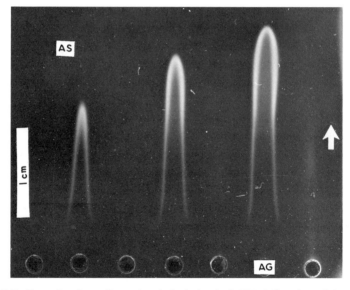

FIGURE 8. Example of one-dimensional single (rocket) EID. Migration of the antigen (bovine serum albumin) was upward toward the anode into antiserum-charged gel. The lancets of precipitate formed were progressively higher as antigen concentration was increased from left to right. Only the leftmost lancet came to equilibrium with antibodies, as indicated by the sharpness of its tip. For greatest quantitative accuracy, electrophoresis would have had to be continued until the center and right lancets had become equally sharp-tipped. This microtest was done on a double-width microscope slide using 0.6-μl volumes of antigen and 0.1 ml of antiserum, with electrophoresis lasting 1 hr.

the antiserum-charged agarose, this problem can be nullified and much improved qualitative as well as quantitative information obtained (Figure 9). The result is two-dimensional single EID, which can be as pleasing estheti-cally as scientifically (Figure 10). As in one-dimensional EID, each loop of precipitate that is formed indicates by its height and other characteristics the relative concentrations of antigen and antibodies forming it (cf. Crowle, 1973, Chapter 6). Consequently, by using this technique one can study

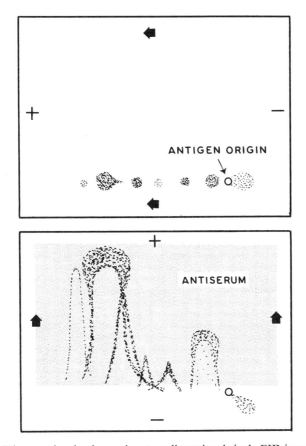

FIGURE 9. Diagram showing how micro-two-dimensional single EID is performed on a double-width microscope slide. After the antigen solution (e.g., human serum) has been electrophoresed at 10 V/cm along one edge of the slide through 1.5% agarose for 1 hr in pH 8.6 barbital–acetate buffer of ionic strength 0.05, antiserum is allowed to diffuse into the shaded area, lower diagram, and electrophoresis is repeated at 5 V/cm for a longer time (2–3 hr, depending on various conditions) across the width of the slide. Each antigen forms its own independent loop of precipitate. Only an illustrative few loops have been drawn here in order to keep the diagram simple.

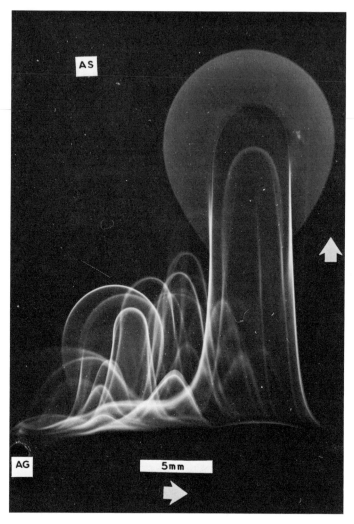

FIGURE 10. An example of micro-two-dimensional single-EID analysis, human serum antigens being reacted with polyvalent horse antiserum to human serum. Note the dimensions of this test. The volumes of antigen and antiserum used were 0.6 μl and 0.25 ml, respectively.

multiple antigen–antibody systems simultaneously but also individually better than by any other currently known technique.

In double EID, as in double diffusion, both reactants move. As shown in Figure 11, electrophoretic conditions are chosen so that antigen and antibodies move toward each other. For example, since antibodies tend to move

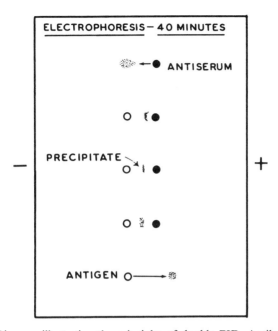

FIGURE 11. Diagram illustrating the principles of double EID. Antibodies from the antiserum migrate electroosmotically toward the cathode, while antigen migrates electrophoretically toward the anode. A small band of precipitate forms where the two cross through each other. As indicated, the test can be completed in a few minutes; 1-μl volumes of both antigen solution and antiserum suffice for performing this test.

toward the cathode at pH 8.2 in agar gel while many antigens move toward the anode under the same conditions, one can set up a double-EID test by cutting two wells in 1% agar gel cast on a microscope slide, charging the anodic well with antiserum and the cathodic well with antigen, and then submitting the slide to electrophoresis for a few minutes. During electrophoresis antigen and antibodies will migrate toward each other rapidly and precipitate much as they do in a double-diffusion test.

Two outstanding advantages to this test are extreme rapidity (results in a few minutes) and good sensitivity. The latter is due to reacting total amounts of antigen and antibody applied to the test plate with essentially no loss by diffusion. This is in contrast with the typical double-diffusion plate test in which most of each reactant diffuses radically, and uselessly, away from the reaction area between their respective sources. However, a major disadvantage to double EID is that unless the reactants are relatively well balanced with each other serologically they may migrate through each other too fast to precipitate well. By comparison, in a double-diffusion plate such imbalance is much less important because the antigen–antibody reaction develops much more slowly and excesses of one reactant over the other tend to be self-correcting because the stronger reactant dilutes itself faster by diffusing faster. Despite its inherent disadvantage of sensitivity to serologic imbalance, and because of its speed and sensitivity, the double-EID test (also known as counter immunoelectrophoresis, immunoosmophoresis, electrosyneresis, and a number of other names) has become a well accepted and widely used analytic technique. Its most frequent application currently is to screen blood donors for hepatitis virus.

A two-dimensional double-EID test has been invented (cf. Crowle, 1973, Chapter 6), but I shall not describe it here because it seems unlikely that it will be as useful for protein analyses as two-dimensional single EID. However, there are some other still little-used immunodiffusion techniques that do merit brief description here because they most certainly will become powerful and welcome additions to the protein chemist's repertoire of analytic techniques once he knows about them.

A first attempt at combining electrofocusing with immunodiffusion in immunoelectrofocusing seemed unpromising because antigen resolution was not better than that seen in immunoelectrophoresis (Riley and Coleman, 1968). However, replacing agarose gel with polyacrylamide for the primary electrophoresis and using electrophoresis instead of diffusion for reacting electrophoresed antigens with their antibodies has produced what seems to be an exceedingly powerful compound technique which readily detects minute electrophoretic variations in one kind of antigen which hitherto have been obscure or unsuspected (Jerka and Blanický, 1973; Skude and Jeppsson 1972).

Even without electrofocusing, primary electrophoresis in polyacrylamide by conventional disk-electrophoresis methods can yield startlingly high resolution of antigen isomers, providing an appropriate immunodiffusion technique for detecting the separated antigens is employed. The trick is to detect these antigens before their resolution is lost by their diffusing into each other, for when this occurs, separation resembles conventional separation in agarose (Catsimpoolas, 1969). One way in which this has been done has been called "immunocore electrophoresis" and consists of casting a rod or core of antiserum-charged agarose into the hollow of a cylinder of polyacrylamide gel in which electrophoresis has just been completed (Zeineh et al., 1973). Antigen–antibody precipitation quickly develops at the edge of each separated disk of antigen at the polyacrylamide–agarose interface, and apparently also between such disks at the periphery of the polyacrylamide in such a way that their resolution is comparable to that seen in conventional polyacrylamide disk electrophoresis, but with the advantage that disks are specifically identified by their reactions with antibodies.

Two-dimensional EID taking advantage of starch gel for primary electrophoresis readily identifies isomers of serum antigens; the technique is routinely used to detect risk of developing emphysema by serum α_1-trypsin inhibitor phenotyping (Kueppers and Black, 1974). Similar analyses can be performed using polyacrylamide in place of starch to study the microheterogeneity of such serum proteins as the haptoglobins (Grubb, 1973).

The combination of chromatography with immunodiffusion has not been well explored. Serum antigens can be separated in Sephadex G-200 thin-layer chromatography and reacted with antiserum; the result resembles an immunoelectropherogram, except that antigens have been fractionated according to molecular size rather than electrophoretic charge (Grant and Everall, 1966). Chromatoimmunodiffusion, as one might call this technique to avoid confusing it with "immunochromatography," which has come to be used to refer to other, unrelated, techniques, deserves more attention and development. Potentially especially interesting would be combining thin-layer chromatography with electroimmunodiffusion in an analog of two-dimensional single EID, a compound technique which so far as I know has not yet been tried.

Two advantages that double EID offers over immunodiffusion are speed and increased sensitivity, both accomplished by moving the reactants against each other electrophoretically. Frequently, the reactants have characteristics which are similar enough to make electroimmunodiffusing them difficult. There are other disadvantages to EID, including requirement of electrophoresis apparatus and of buffers of composition which may not readily support antigen–antibody precipitation. An alternative technique for moving the reactants against each other which does not require electrophoresis and which can be used with buffers of nearly any composition is known as immunorheophoresis (Van Oss and Bronson, 1969). Its principle is to make water evaporate rapidly from a zone of gel between sources of antigen and antiserum and thus to draw the reactants together hydrodynamically. Although hitherto rarely used, this technique seems worth knowing as an alternative to those more commonly employed.

V. HELPFUL HINTS

I have written little about technical details here because this information usually can be obtained from the literature once a technique has been selected from general principles and information that I have outlined. However, I should like to mention two techniques which the reader will find useful whatever method for immunodiffusion he selects. One helps solve the universal problem of recording and storing results from immunodiffusion tests; the other helps solve another universal problem of preparing monospecific antiserum.

A most convenient way of recording immunodiffusion test results is to make contact prints of the stained immunodiffusogram on Diazochrome

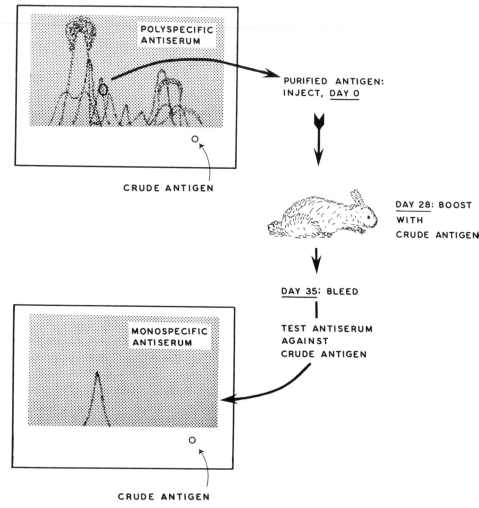

FIGURE 12. A diagram showing how to use preparatory EID to obtain minute quantities of purified antigen with which to prime a rabbit for selective anamnestic response to this same antigen when it is injected later in much larger quantities as an unpurified mixture of antigens. The purpose of this technique is to make monospecific antiserum against native antigen which has been purified from a complex mixture of antigens by direct immunochemical rather than empirical biophysical techniques.

film (Crowle, 1973, pp. 198–199). This film can be handled in room light but is exposed by a few seconds of intense lighting, such as provided by a photoflood light. It is developed in one minute with 10% ammonium hydroxide and then is blotted dry. The product is a flexible, durable, positive, faithful reproduction of the original which can be stapled or taped into a data book. Diazochrome is sensitive only to blue light; consequently, the immunodiffusion pattern must be stained some other color to record well.

Monospecific antisera can be prepared against many antigens by taking advantage of two-dimensional single EID to prepare minute amounts of antigen quickly, directly, and with minimal denaturation, and of the immunologic phenomenon of specific anamnesis (Crowle, *et al.*, 1972). For intance, from the pattern obtained on two-dimensional single EID of a mixture of antigens, one selects the loop of precipitate representing the antigen for which monospecific antiserum is to be made and then cuts a segment of precipitate forming this loop free of the surrounding gel (Figure 12). This, and perhaps additional segments from two or three more electroimmunodiffusograms, are leached free of unreacted proteins, emulsified in Freund adjuvant, and injected into a rabbit. This injection primes the rabbit in an immunologically specific way so that if it is injected three or four weeks later with large amounts of the crude antigen mixture it will begin producing useful quantities of precipitins, solely to the antigen originally injected, among all the antigens in the mixture. Thus, by priming this rabbit with only the minute amount of antigen readily and directly prepared by EID and then boosting it with the much larger amounts of the same antigen present in the crude mixture, one can obtain a monospecific antiserum with very little investment of time and effort, and with the guarantee of success which any explicit technique has over empirical methods such as are used for protein purifications in classic biochemistry. Needless to say, monospecific antiserum is an extremely powerful reagent, having many uses other than in immunodiffusion.

Although this chapter has only touched on some fundamentals of immunodiffusion tests and described briefly some of the techniques most likely to be useful to the protein analyst, perhaps it will encourage him to use immunodiffusion tests more frequently by having made them appear to be somewhat more logical and less mysterious than they freqently seem to those only passingly acquainted with them.

VI. REFERENCES

Afonso, F., Affonso, A., and Sanaguire, R. S. (1972) *Clin. Chem. Acta* **41**: 275.
Bjerrum, O. J. and Lundahl, P. (1974) *Biochim. Biophys. Acta* **342**: 69.
Catsimpoolas, N. (1969) *Immunochemistry* **6**: 501.

Crowle, A. J. (1956) *J. Lab. Clin. Med.* **48**: 642.

Crowle, A. J. (1973) *Immunodiffusion*, 2nd ed., Academic Press, New York.

Crowle, A. J. and Atkins, A. A. (1974) Unpublished work.

Crowle, A. J., Revis, G. J., and Jarrett, K. (1972) *Immunol. Commun.* **1**: 325.

Grabar, P. and Williams, C. A., Jr. (1953) *Biochim. Biophys. Acta* **10**: 193.

Grant, G. H. and Everall, P. H. (1966) *Protides Biol. Fluids Proc. Colloq.* **13**: 321.

Grubb, A. (1973) *Protides Biol. Fluids Proc. Colloq.* **21**: 649.

Jerka, M. and Blanický, P. (1973) *Biochim. Biophys. Acta* **295**: 1.

Krøll, J. (1970) *Protides Biol. Fluids Proc. Colloq.* **17**: 529.

Kueppers, F. and Black, L. F. (1974) *Am. Rev. Respir. Dis.* **110**: 176.

Ouchterlony, Ö. (1948) *Arkh. Kemi Mineral. Geol.* **26**B: 1.

Oudin, J. (1946) *C. R. Acad. Sci., Paris* **222**: 115.

Ressler, N. (1960) *Clin. Chim. Acta* **5**: 359.

Riley, R. F. and Coleman, M. K. (1968) *J. Lab. Clin. Med.* **72**: 714.

Skude, G. and Jeppsson, J. -O. (1972) *Scand. J. Clin. Lab. Invest.* **29** (*Suppl.* 124): 55.

Van Oss, C. J. and Bronson, P. M. (1969) *Immunochemistry* **6**: 775.

Williams, C. A., Jr. and Chase, M. (1971) *Methods in Immunology and Immunochemistry*, Vol. 3, Academic Press, New York.

Zeineh, R. A., Mbawa, E., Pillay, V. K. G., Fiorella, B. J., and Dunea, G. (1973) *Biochim. Biophys. Acta* **315**: 1.

ISOELECTRIC FOCUSING IN POLYACRYLAMIDE GEL

4

JAMES W. DRYSDALE

I. INTRODUCTION

From many viewpoints, isoelectric focusing (IEF) represents a major advance in methodology for high-resolution separation of proteins and other amphoteric macromolecules. IEF is essentially an equilibrium method for segregating amphoteric molecules by electrophoresis in stable pH gradients according to their isoelectric points (pI). In only one or two experiments it is theoretically possible to display all components whose pI values differ by as little as 0.01 pH units. Such exquisite resolution by charge is not normally obtainable by other procedures such as electrophoresis or ion-exchange chromatography. Because of this built-in resolution, IEF is a more definitive technique for examining charge heterogeneity and is, therefore, a rigorous test of homogeneity. In addition, because substances are also concentrated as they are separated by IEF, the technique lends itself to both analytical and preparative purposes. Finally, IEF defines an important parameter, the isoelectric point (pI), which gives information not only about the composition and conformation of macromolecules but also allows a rational approach to further experimental manipulations.

JAMES W. DRYSDALE, Department of Biochemistry and Pharmacology, Tufts University School of Medicine, Boston, Massachusetts 02111.

As originally developed, IEF was primarily a preparative technique. Fractionations were performed in sucrose density gradients which served as an anticonvective medium to support the pH gradient and focused protein zones. However, such systems were rather expensive in both time and materials. They were also subject to many practical problems arising from excessive diffusion and zone instability and were not readily adaptable for routine analytical procedures. More recently, other anticonvective media have been tested. This chapter will describe some aspects of IEF in polyacrylamide gel, a medium which realizes much of the potential inherent in the principle of IEF and which also offers considerable experimental flexibility. For reviews of theoretical and practical aspects of IEF see Svensson (1961, 1962), Vesterberg (1970), Haglund (1971), and Rilbe (1973). For early developments in gel electrofocusing see Awdeh *et al.* (1968), Hayes and Wellner (1969), Wrigley (1968), Dale and Latner (1969), Catsimpoolas (1971), Williamson (1973), and Righetti and Drysdale (1971, 1974).

II. BACKGROUND

The principle of IEF is outlined in Figure 1. A stable pH gradient increasing progressively from anode to cathode is established by electrolysis of carrier ampholytes in a suitable anticonvective liquid medium. Strong acid and base are used as electrolytes to prevent electrolytic decomposition

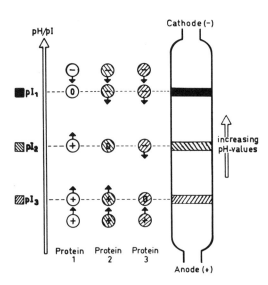

FIGURE 1. Separation of proteins by IEF. Three proteins with isoelectric points, pI_1, pI_2, and pI_3, schematically electrofocused in a column. From Haglund (1971) courtesy of LKB Instruments, Sweden.

of the ampholytes. A protein or other amphoteric macromolecule introduced into the pH gradient below its pI will be positively charged and be repelled by the anode. As it moves into regions of higher pH it will lose positive charge and gain negative charge, e.g., through deprotonation of carboxyl or amino functions. Eventually it will reach a pH region at which its net electrical charge is zero (pI). Should it diffuse away from its pI, it will develop a net charge and be repelled back toward its pI. Thus, by counteracting backdiffusion with an appropriate electrical field, the protein will reach an equilibrium position and so be concentrated, or focused, at its pI. By this means different proteins can be segregated into narrow zones in the same pH gradient.

As might be expected, the degree of separation of two ampholytes is a function of the range of the pH gradient in which they are focused. The narrower the pH gradient, the better is the separation (Figure 2).

Although the principle of IEF has been known and used since the early 1900s, it is only recently that systems have been sufficiently refined for routine practical purposes. One of the major obstacles in early experiments was the lack of suitable carrier ampholytes for developing smooth and sufficiently stable pH gradients to allow equilibrium focusing. This problem was largely solved through the efforts of Kolin, Svensson (now called Rilbe), and Vesterberg. Kolin pointed out the importance of developing uniform field strengths with electrolytes of a high buffering capacity and of stabilizing the pH gradient against convective mixing (Kolin, 1954, 1955). However, he was limited by the availability of suitable ampholytes and could not establish sufficiently stable pH gradients (Kolin, 1958). The next major advance came from the work of Svensson (Svensson, 1961, 1962; Rilbe, 1973), who laid

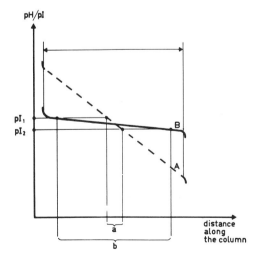

FIGURE 2. The distance between two zones pI_1 and pI_2 will be greater when the pH range is made narrower. Line A represents the pH gradient for the wider pH range and line B the narrower pH range. From Haglund (1971) courtesy of LKB Instruments, Sweden.

the theoretical foundations for present day systems of IEF. Svensson specified the properties required for suitable carrier ampholytes for IEF. He pointed out that ampholytes should have a good buffering capacity, conductivity and solubility at their pI, closely spaced pI and pKa values, and be readily distinguished and separated from proteins. Many of these requirements were met by the synthesis of carrier ampholytes by Vesterberg (1969). This preparation paved the way for the practical application of IEF. The ampholytes were formed by reacting acrylic acid and polyethylene polyamines to give a series of homologs and isomers substituted to different extents by carboxyl and amino functions. The general reaction can be illustrated by

$$R_1\overset{+}{-}NH_2-(CH_2)_2\overset{+}{-}NH_2-R_2 + CH_2 = CH-COO^- \rightleftharpoons$$

$$R_1\overset{+}{-}NH_2-(CH_2)_2\overset{+}{-}NH-R_2$$
$$\mid$$
$$CH_2-CH_2-COO^-$$

A large number of ampholytes with pI values between pH 3 and 11 are generated by this synthesis. This pH range encompasses the pI values of most

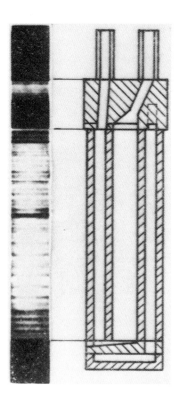

FIGURE 3. Photograph of cell after IEF of carrier ampholytes. The striations are due to strong refractive-index gradients caused by focused ampholytes. From Rilbe (1970) by permission of the New York Academy of Sciences.

proteins. These ampholytes range in size from 300 to 1000 daltons, although about 1% may be as large as 1000–5000 daltons (Gasparic and Rosengreen, 1974). The exact number formed is not known. Estimates of focused ampholytes detected by birefingence (Rilbe, 1970) (Figure 3), UV absorption, caramelization with sugars (Felgenhauer and Pak, 1973) or staining (Otavsky and Drysdale, unpublished data) indicate that there may be about 50–100 different species in the pH range 3–11. The wide-pH range ampholytes may be subfractionated by IEF to give preparations capable of forming pH gradients of less than 1 pH unit. The advantages of narrow pH ranges for improved separation of proteins of similar pI has been discussed earlier. Ampholytes (Ampholines) prepared according to Vesterberg's synthesis are now available commercially from LKB Produkter AB, Bromma, Sweden. However, all of the required reagents are commercially available and the entire process may be readily accomplished in the laboratory.

Although such ampholytes meet many of Svensson's criteria, they have several undesirable features. For example, although their average UV absorbancy is small, that of individual focused ampholytes may be sufficient to mask or obscure focused protein zones (Figure 4). In addition, the ampholytes also react with many commonly used protein stains. Happily, this problem has now been circumvented and several sensitive methods have been described for direct staining of proteins. Perhaps the most troublesome

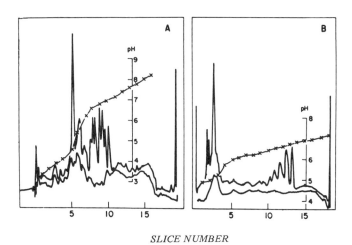

SLICE NUMBER

FIGURE 4. Ampholine interference in gel scans at 280 nm. Approximately 50 μg of protein (myeloma serum) was fractionated in the pH ranges (A) 3–10 and (B) 5–7. After focusing the gels were scanned at 280 nm (upper tracing). Replicate gels containing the same amount of Ampholine but no protein were run in parallel. Note the large, but uneven, adsorbancy due to separated ampholytes in these gels (lower tracing). From Righetti and Drysdale (1971) by permission of Elsevier Press.

aspect of present-day ampholytes is their uneven distribution in certain pH ranges. Figure 5 shows the conductance profiles and UV absorbancies of three Ampholine preparations. Ampholines in the pH ranges 3–6 and 7–10 form smooth pH gradients with a fairly even conductance course. The conductance in the pH 5–8 gradient is, however, quite uneven. Consequently, local hot spots may develop and give rise to convective mixing and risk of protein denaturation under the high voltages required for optimum resolution.

The early experiments with IEF in sucrose density gradients clearly demonstrated the considerable advantages of the technique in high-resolution separations of proteins. However, IEF in all liquid media has serious disadvantages that severely limit its applicability. For example, isoelectric precipitation of focused zones often restricts the amount of applied sample to the extent that minor components escape detection. Diffusion and mixing of samples during elution of the sucrose density gradient after focusing can

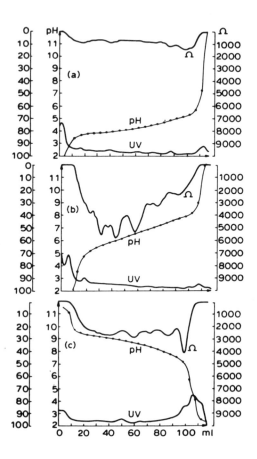

FIGURE 5. Conductance, pH-gradient, and UV-absorbance profiles of three Ampholine pH ranges. The conductance course of the pH gradients pH 3–6 (a) and (c) 7–10 are very even. The pH gradient 5–8 (b) contains low-conductivity ampholytes around pH 7 and shows a more uneven conductance course. This gradient is consequently more sensitive to thermal convection caused by Joule heating in the low-conductance region. Generally, the conductance course is more even, when shorter pH regions are used. From Davies (1970) courtesy of LKB Instruments, Sweden.

greatly diminish the resolution actually achieved in the column (Catsimpoolas, 1973*a,b*) (see also section V.D). In addition, the subsequent analysis of multiple samples with variable interference from ampholytes is tedious. These latter problems have been elegantly resolved by Catsimpoolas with arrangements for scanning sucrose density gradients under voltage (see Chapter 2). However, this procedure calls for complex equipment and is not very convenient for routine analysis of multiple samples.

A variety of alternative anticonvective media such as paper, cellulose, acetate membranes, agarose or polyacrylamide gels, or Sephadex beds have been explored. Although these substances are extensively used for electrophoresis, many of them are unsuitable for IEF, largely because of an excessive electroendosmosis. At present, the two most suitable media seem to be polyacrylamide gel and Sephadex beds (Righetti and Drysdale, 1974). Both have been adapted for routine analytical and preparative IEF. Polyacrylamide gel has, of course, many desirable properties for electrophoretic techniques. It is an excellent anticonvective medium, is chemically inert, optically clear, and has a satisfactory mechanical strength. More importantly for IEF, it has a low electroendosmotic effect. This review will deal exclusively with IEF in polyacrylamide gel.

Several methods for IEF in polyacrylamide gel (GEF) were published separately and almost simultaneously between 1968 and 1970 (Righetti and Drysdale, 1974). These early studies dramatized the improved resolution and experimental flexibility obtainable by IEF in this medium. As in electrophoresis, two basic systems emerged for GEF, individual cylinders run in parallel and thin rectangular slabs with parallel tracks. The choice between cylinders and slabs depends on the particular requirements. Slabs allow excellent comparisons of similar samples. They also permit sample application as anodic or cathodic species without risk of exposure to extremes of pH at the electrodes. Gel cylinders, on the other hand, allow greater experimental flexibility and also permit analyses of more dilute samples. Focusing in cylinders also offers simpler quantitation by densitometry and the use of other analytical techniques deployed after gel electrophoresis. Both systems can be adapted for the simultaneous fractionation of multiple samples in the same or different pH gradients.

III. METHODOLOGY

A. Apparatus

Three items are required for GEF: an electrofocusing cell, a high-voltage power supply, and a means of efficiently cooling the gels.

1. Electrofocusing Cell

Best results with GEF in gel cylinders or slabs are given with apparatus designed specifically for GEF. Suitable apparatus for both is readily made or is available commercially. Although both the tube and the slab systems bear certain similarities to apparatus for electrophoresis, there are major design differences that are important for reliable and reproducible results by GEF. First, by necessity most electrophoresis apparatus have large electrolyte volumes for buffering changes in pH during electrolysis. When used for GEF, these chambers may create excessive convective mixing in the buffer compartment and so disrupt the connecting pH gradient between the electrodes and the gel. Second, considerable resistance (up to 4 kΩ) may develop across a small analytical gel. This resistance is not uniformly distributed in the gel and local hot spots may arise in areas where the conductance is uneven. This heat must be efficiently dissipated to reduce convective mixing and prevent heat denaturation of proteins. Unfortunately, most electrophoretic apparatus do not allow sufficient gel cooling. Both of these factors should therefore be considered in the design of GEF equipment. Convective mixing in electrode chambers may be reduced by minimizing electrolyte volumes and the distance between electrode and gel tube. In most slab apparatus, the electrodes are placed directly on the gel (Awdeh et al., 1968). Cooling in both tube and slab systems is usually achieved by circulating coolant or by direct contact with a cold block.

2. Gel Cylinders

Righetti and Drysdale (1971) developed an apparatus for analytical and preparative GEF in cylinders of polyacrylamide. The salient features include efficient cooling of gel tubes by circulating antifreeze at −5°C to +4°C and small electrolyte chambers with electrodes in close juxtaposition with the tube ends. The apparatus can hold up to 12 gels (10 × 0.3 cm i.d.). Gels (approximately 1-ml volume) are usually cast in plastic tubes from which they are readily extruded. Quartz tubes may also be used interchangeably for direct scanning in the UV. Figure 6 shows analytical and preparative models of apparatus with these features.

3. Gel Slabs

Several suitable systems have been described for GEF in thin slabs of polyacrylamide. Awdeh et al. (1968) described a simple system in which

FIGURE 6. Analytical and preparative apparatus for GEF in gel cylinders showing fractionations of hemoglobin variants. The preparative apparatus also contains a fractionation of 50 mg of human placental lactogen. The multiple forms of this hormone are evident when viewed against the dark background. Courtesy of MRA Corp., Boston, Mass.

thin slabs of gel were polymerized between vertical glass plates. One face of gel was exposed for sample application. Electrode contact was made by placing wetted carbon electrodes directly on the gel surface. No electrolyte vessels were required. Similar systems are now widely used. In some, a template is used to imprint indentations in the polymerizing gel to facilitate sample application (Figure 7). Most systems also now employ platinum electrodes which make gel contact through suitably impregnated paper or felt wicks.

B. Electrolyte Solutions

The choice of electrolyte solutions is dictated by the need to maintain good conductivity for the formation of the pH gradient. Obviously the pH of the acidic and basic electrolytes should encompass the pH gradient to be developed in the gel. In the slab systems, where samples are usually applied some distance from the electrodes and electrical contact is made directly through narrow wicks, fairly strong acids and alkalis, e.g., 1 M phosphoric acid and 1 M sodium hydroxide, are commonly used. However, in the tube system, weaker solutions, e.g., 10 mM, are preferred to minimize the risk of

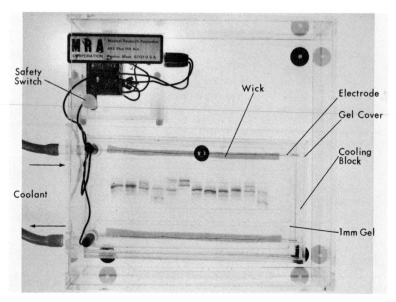

FIGURE 7. Apparatus for GEF in thin slabs of polyacrylamide gel with separation of hemoglobin variants. Courtesy of Medical Research Apparatus Corp., Boston, Mass.

exposing samples to damaging pH. In this case, samples should be adequately buffered prior to application.

C. Gel Composition

Gels for GEF by cylinder or slab are usually prepared by incorporating the appropriate ampholytes into the gel before polymerization. With slabs, however, the ampholytes may be diffused into preformed gels to prevent their possible modification during gel polymerization. During diffusion, free radicals remaining after polymerization will be removed. Polymerization may be initiated by persulfate or riboflavin or both. N,N-Methylene (bisacrylamide) is commonly used as a cross-linker, though diallyl tartaramide may be used to facilitate gel dissolution after focusing. Both the acrylamide and the cross-linker should be recrystallized or be of the highest quality. Impurities in the acrylamide may cause troublesome electroendosmosis with attendant problems of gel thinning and pH-gradient instability. Other important general considerations in the composition and formation of polyacrylamide gels have been discussed by Chrambach and Rodbard (1972).

Because GEF is an equilibrium method, fairly wide variation in gel composition may often be tolerated. However, it should be noted that meaningful results with GEF can only be obtained if all components reach their pI in the time allotted for the experiment—with the minimum trauma to the samples. Unfortunately the same anticonvective properties of polyacrylamide gel that give high-resolution banding patterns introduce a potentially troublesome factor, molecular sieving, that cannot be ignored. Although molecular sieving is often advantageous in gel electrophoresis, it should ideally be minimized in GEF, particularly when attempting to focus large proteins. Consequently, it is advisable to use the highest porosity gel consistent with satisfactory mechanical strength. The porosity of polyacrylamide gel is largely determined by the relative amounts of acrylamide and cross-linker (Hjertén, 1972; Blatter et al., 1972). For most purposes, gels containing 4% acrylamide and 0.16% N,N'-methylene (bisacrylamide) will, under suitable electrolysis periods, allow equilibrium focusing of proteins of 500,000 daltons in a 10-cm gel in about 6–8 hr. For smaller proteins (less than 200,000 daltons) more robust gels containing 5% or 6% acrylamide may be preferred. For larger proteins, lower gel concentrations and/or longer electrolysis periods should be used. Rodbard et al. (1972) have described gels containing very high levels of cross-linker, e.g., 25% of the acrylamide concentration, in which molecules up to 25×10^6 daltons can be focused. These gels are, however, rather brittle and are not optically clear.

It follows from the above that another important requirement for GEF is the development of sufficiently stable pH gradients to permit proteins to overcome sieving effects and reach their equilibrium positions. In addition to suitable apparatus, several other factors such as ampholyte concentration and electrolysis conditions are important in this regard. Failure to develop stable and reproducible pH gradients leads to considerable difficulties in interpretation of banding patterns. This point is demonstrated in Figure 8, which shows banding patterns of the same sample in gels containing different amounts (1% and 4% wt./vol.) of the same Ampholine preparation. The pH gradient in the gel with 4% Ampholine was smooth and almost linear over the length of the gel. By contrast, the gel with only 1% Ampholine gave an irregular pH gradient. As a consequence, the banding patterns appeared quite different, at least in the position of the proteins. The correspondence of the major species in each gel could only be confirmed by characterization according to measured pI. Figure 9 shows typical effects of electrolysis time and ampholyte concentration on the course of pH gradients developed in GEF. Similar effects have been found by Finlayson and Chrambach (1971). In general, a minimum level of 1.5% (wt./vol.) of most Ampholines is recommended for GEF. This dependency on Ampholine concentration is probably not just a consequence of increased viscosity in stabilizing the pH

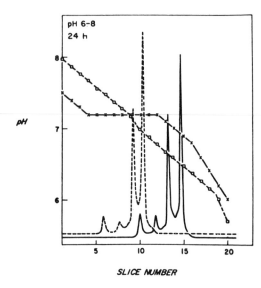

SLICE NUMBER

FIGURE 8. Fractionation of hemoglobins from a rat hemolysate in the pH 6–8 range at different ampholyte concentrations. Samples of hemolysate (100 μg protein) were fractionated in 4% acrylamide gels containing either 1 or 4% Ampholine LKB 8154, pH 6–8. After a total electrolysis period of 24 hr, the gels were scanned at 576 nm in a recording spectrometer [(—) 1% ampholyte, (– -) 4% ampholyte] and subsequently sectioned as before for a determination of the pH gradient [(×–×) 1% ampholyte, (□—□) 4% ampholyte]. From Righetti and Drysdale (1971) by permission of Elsevier Press.

gradient as may occur with sucrose or glycerol. A more reasonable explanation is that a certain ampholyte level is required to maintain good electrical conductivity throughout the gel and prevent hot spots from local irregularities in conductivity.

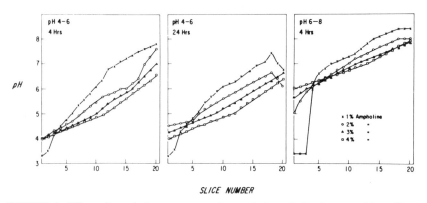

SLICE NUMBER

FIGURE 9. Effect of ampholyte concentration and electrolysis time on pH gradients. The 4% polyacrylamide gels containing 1, 2, 3, or 4% ampholytes (LKB 815, pH 4–6; LKB 8154, pH 6–8) were subjected to electrolysis for the times indicated. The gels were then fractionated into 20 slices. Each slice was eluted with 0.5 ml water, and the pH of the gel eluates was measured with a combination microelectrode. From Righetti and Drysdale (1971) by permission of Elsevier Press.

D. Electrolysis Conditions

Generally, more rapid fractionations and sharper resolution are obtained at high potential differences across the gel.

Svensson (1961) showed the distribution of an amphoteric molecule in IEF is determined by the relationship $Cui/qk = D(dC/dx)$, where $C =$ the concentration of component and ion constituent in arbitrary mass units per unit volume, $u =$ electrical mobility in $cm^2 V^{-1} sec^{-1}$ of ion constituent except H^+ and OH^-, $i =$ electrical current in A, $q =$ cross-sectional area in cm^2 of electrolytic medium measured perpendicularly to the direction of current, $k =$ conductance of medium in $\Omega^{-1} cm^{-1}$, $D =$ diffusion coefficient in $cm^2 sec^{-1}$ of component corresponding to the ion constituent with mobility μ, and $x =$ coordinate along the direction of current. If the pH gradient and conductivity are constant throughout a focused protein zone, the protein will have a Gaussian distribution with inflection points at $x_1 = \pm(qkD/pi)^{1/2}$, where $p = -du/dx$ and x_1 denotes the width of the focused zone measured from the top of its distribution to the inflection point (one standard deviation). For further theoretical considerations of isoelectric fractionations of proteins and ampholytes see Svensson (1961, 1973).

The degree of focusing is directly proportional to the square root of the field strength $E = i/qk$. Also the most rapid electrofocusing is obtained by applying the maximum potential difference without adverse heating effects. However, because of the wide variations in conductivity in the gels, electrolysis conditions must be carefully monitored to prevent undue heating, sample denaturation, convective mixing, or pH-gradient instability. The amount of heat generated in an electrofocusing cell (joule heat per cm^3 of medium) is given by the expression $W = i/q^2k$. In the early stages of GEF, the ampholytes distributed throughout the gel have a high conductivity. However, as they migrate toward their pI the conductivity decreases to reach a low plateau when the pH gradient has stabilized. Since the heating effect is related to both the applied potential difference and the current, it is necessary to regulate the power delivered to the cell. This is usually achieved by carefully monitoring the current and the voltage during the development of the pH gradient and regulating either. However, it is now possible to automatically regulate the power delivered to the cell by regulating the pulse rate of a predetermined potential difference. Decreasing conductivities are automatically matched by increasing pulse frequency or width until, at equilibrium, the system is controlled only by the applied potential difference (Drysdale, 1974).

In practice, with cylindrical 3-mm thick gels cooled at about 4°C, the wattage (volts × amps) should not exceed 0.2 W/gel. This may be arranged by running gels initially at a constant current of 0.5 mA/gel until the voltage required to drive this current increases to 300 or 400 V. The voltage may then be stabilized at this level when the current will decline to a constant plateau after the pH gradient has stabilized. With regulated pulse power supplies, an average power of 0.2 W/gel at a potential difference of 400 V gives satisfactory results.

With most slab apparatus in which gels are more efficiently cooled higher power settings may be used. In some instances, an average power of 50 W at a voltage of 1000 V has been used to give analyses in about 1 hr (Wadström, 1974). However, the pH gradients tend to be unstable. We routinely use a regulated power of 2 W/gel (1 mm thick) .Under these conditions, hemoglobin (mol. wt. 68,000) will focus in 4% gels in about 4 hr, whereas ferritin (mol. wt. 450,000) requires at least 8 hr.

E. Sample Application

1. In Cylinders

Samples may be applied in several ways, either in a small volume from concentrated solution or throughout the gel from dilute solution. When focusing in gel cylinders, samples are usually applied in a fairly concentrated form on top of preformed gels. Since most proteins are usually kept in solution in their anionic form, it is generally preferable to apply samples at the cathodic end to facilitate their rapid entry into the gel. On the other hand, if proteins are to be applied in their cationic form, the electrodes and electrolytes should be reversed. Samples should be applied in a dense buffered solution such as 1% ampholyte in 10% glycerol under a protective layer of ampholyte to prevent contact with electrolytes. Although IEF occurs in a saltfree medium, small amounts of salt, e.g., up to 50 mM, may be added in the buffering solution. The volume of sample is not critical. Multiple applications of dilute samples may be made at intervals, provided sufficient time is subsequently given to ensure that components from the last application reach their equilibrium positions. Samples may also be incorporated directly into the gel before polymerization if they are not affected by gel constituents. In this case, the total amount of salt added in the sample should be minimized so that the salt does not carry too much of the current.

2. In Slabs

Samples may be applied anywhere on the surface of the slab in parallel tracks. Consequently, there is less likelihood of modification by adverse pH of electrolytes. This method also offers a ready means of determining equilibrium focusing by demonstrating coalescence of samples applied at different positions in the gel. Samples may be applied on strips of suitable absorbants (Wadström and Smyth, 1973) or in small indentations in the gel (Leaback and Rutter, 1968).

F. Sample Load

Best resolution is usually given at low sample levels on the gel. Consequently, for most analytical procedures in 3-mm cylinders or 1- to 2-mm slabs, a level of 1–5 μg per focused band gives satisfactory staining patterns with dyes such as Coomassie Brilliant Blue. However, for detection of trace components, gels may be overloaded with major components. The maximum amount of protein that can be focused as a discrete band depends on the solubility at its pI, the course of the pH gradient, and the proximity of other proteins. With many proteins, band densities of up to 800 μg/band/cm^2 of gel are possible. Indeed, it is often possible to apply such large amounts that colorless proteins become clearly visible as opaque regions when focused in the gel (see Figure 6).

G. Measurement of pH Gradients

The pH gradients developed in the gel may be measured in several ways. This measurement is of utmost importance in characterizing banding patterns in different experiments. Although reproducibility of pH gradients developed with any given Ampholine under standard conditions may be excellent, batch variability and differences in experimental procedure may give slightly different results. Such variability can, however, be readily corrected by correlating banding patterns with the actual pH gradient developed in the gel.

Measurement of pH gradients also provides a direct estimate of the pI of individual components. This value will generally correspond to the isoionic point which, strictly defined, is that pH which does not change on addition of a small amount of pure protein to the solution (Cannan, 1942). This definition also pertains to a focused protein zone, where the buffering is determined by the carrier ampholyte. It does not, however, apply to focused carrier ampholytes since the pH of the ampholyte zone is dependent on the concentration of carrier ampholyte.

Several factors can affect the measured pI of isoelectrically focused proteins, and it is important to standardize procedures to ensure good reproducibility. Since the dissociation of proteins and other amphoteric molecules is temperature dependent, it is usually desirable to focus at constant temperatures and to estimate the pI at that same temperature. Since many proteins are temperature sensitive, focusing and pI determinations are usually carried out at 0–10°C. However, most estimates of pI by other means are obtained closer to 25°C. It is therefore fortunate that estimates of pI from solutions obtained by IEF at 4°C but measured at 25°C give the same values as when IEF is conducted at 25°C (Vesterberg and Svensson, 1966).

In addition to temperature, other factors may affect the measured pI. For example, the presence of other solutes, such as sucrose, may reduce the dielectric constant of the solution and alter the degree of dissociation of ionizable functions. It should also be borne in mind that solutes such as sucrose or urea may substantially alter the conformation of proteins or other amphoteric macromolecules and so alter the apparent pI. Such shifts in pI have indeed been useful in conformational studies (Ui, 1973).

The pH gradient in gels may be measured in gel eluates with a combination microelectrode. As a general rule, the volume of eluate should not be more than five times that of the gel section to avoid excessive dilution of the ampholytes. Distilled water may be used to elute the ampholytes, but it is often advisable to add low levels of salt (approximately 10 mM) for adequate conductivity. In addition, care must be taken to minimize absorption of atmospheric CO_2. This may be achieved by boiling or by flushing with nitrogen.

Alternatively and more conveniently, pH gradients in cylinders or slabs may be measured directly from the gel with a flat-membrane electrode with a fine tip (e.g., type LOT 403-30-M8, Ingold, Zurich, Switzerland) or by a spear-tip electrode. A particularly suitable model with a 1-mm tip is also available from Ingold Electrodes (Lexington, Massachusetts).

H. pH-Gradient Instability

Several investigators have observed a progressive shift in the pH gradient in gels after prolonged periods of electrolysis. This instability is usually indicated by a progressive flattening of the gradient in the middle regions of the gel and a slow shift toward the cathode. The reason for this phenomenon has not yet been established. With most proteins it is relatively unimportant since equilibrium focusing is usually achieved long before the decay in the pH gradient. However, the problem becomes acute when examining high molecular weight proteins whose mobility may be considerably retarded by the sieving action of the polyacrylamide gel.

Most evidence suggests a multifactorial basis for the pH-gradient instability. Electroendosmosis appears to be a major cause since gradient instability is more pronounced with impure grades of acrylamide gel or in other media with high electroendosmosis, such as Sephadex, agarose, or cellulose acetate. Convective mixing in electrolyte chambers as well as the gels also seems an important factor since the effect is more pronounced in inefficiently cooled apparatus with large electrolyte volumes. Local heating effects caused by uneven distribution of ampholytes and low conductances may also be contributing factors since gradient instability increases with very high potential differences. In this case, much of this problem may be over-come by using sufficiently high levels of ampholytes in the gel (Figure 8) and by carefully regulating electrolysis conditions. However, as pointed out by Haglund (1971), excessive heating and convective mixing may arise in any pH gradient that does not include ampholytes buffering near pH 7. In such cases, a zone of pure water will develop around pH 7. Most of the electrical potential drop will occur at this point and other parts of the gradient will be underfocused. To avoid this, a small amount (e.g., 10%) ampholytes with pH ranges of 3–10 or 6–8 should be added to the narrow-pH-range ampholytes. In situations where pH-gradient instability is appreciable, it is desirable to determine the minimum time required for equilibrium focusing before banding patterns are lost. In this regard, the *in situ* scanning devices developed by Catsimpoolas (1973) are valuable aids.

IV. SAMPLE DETECTION

A. Staining For Proteins

In the early development of gel electrofocusing (GEF), considerable problems were encountered in detecting focused proteins because the Ampholines formed insoluble complexes with many protein stains. This problem was initially overcome by first precipitating focused proteins in acid and then eluting the ampholytes with exhaustive washing—a laborious and time-consuming process. Happily, several direct staining methods with high sensitivity have recently been developed. The success of the methods appears to be in discriminating between complexes of dyes with ampholytes and proteins on the basis of their differential solubility in alcoholic solutions and their temperature and pH stability. For high protein levels in the gel staining with Bromophenol Blue (Awdeh *et al.*, 1968) or Fast Green (Riley and Coleman, 1968) in acid–alcohol solutions is recommended. However, since better resolution is usually obtained at relatively low protein levels, staining methods with the more sensitive Coomassie Brilliant Blue series are normally preferred. Spencer and King (1971) described a simple method in which proteins absorbed the dye from a 0.01% solution in 5% trichloroacetic acid, 5% sulfosalicylic acid, and 25% methanol. In this method, the low dye concentration minimizes background staining. Vesterberg (1972) devised a rapid staining method in which proteins are stained with Coomassie Blue at 60°C in an acetic acid–ethanol solution. Malik and Berrie (1972) have described a method using an extract of Coomassie Blue that has given satisfactory results. However, this method may not be generally applicable since the active staining ingredient appears to be a variable contaminant of Coomassie Blue preparations. We have found a method that combines high sensitivity and low background. Focused gels are immersed with shaking for 6 hr in a solution of 0.05% Coomassie Blue and 0.1% cupric sulfate in acetic acid–ethanol-water (10:25:65). Gels are destained for 4 hr in the same solution but containing only 0.01% Coomassie Blue and finally in acetic acid–ethanol–water (10:10:80). Recently, however, we have experienced troublesome backgrounds with these and other methods with Ampholines in the pH range of 6–9. It is not known whether these represent impurities in recent batches of Ampholines or the addition of other ampholyte substances to bolster the notoriously weak pH 5–8 range. Interestingly, ampholytes prepared in the laboratory according to the original LKB patent (Vinogradov et al., 1973) did not stain (Otavsky and Drysdale, unpublished observations). Future developments might include staining with fluorescent dyes such as fluorescamine (Udenfriend et al., 1972) if its decomposition in aqueous media can be retarded and ampholyte interference prevented. Possibly, procedures for prelabeling proteins with suitable chromophores may prove useful.

B. Histochemical Staining

Analytical GEF in cylinders or thin slabs is generally suitable for histochemical detection or enzymes provided their activity is not lost at their p*I*. Most of the methods devised for detecting enzyme activity after gel electrophoresis can usually be adapted for gel electrofocusing. Appropriate steps must, however, be taken to counteract possible adverse effects of pH from ampholytes. Often, it suffices to carry out incubations in strongly buffered solutions or by a brief pretreatment to equilibrate the gel pH (Wadström and Smyth, 1973).

Although IEF is performed in salt-free media, the ampholytes usually help to keep proteins in solution. Indeed, because of their polyvalent nature, they may even afford greater protection of biological activity than inorganic salts (Vesterberg et al., 1967). In some cases, enzyme stabilization may be due to the chelation of inhibitory heavy metals by the ampholytes. On the other hand, metal chelation may inactivate metalloenzymes on IEF. Latner *et al.* (1970) found that alkaline phosphatase was almost completely inactivated by prolonged periods of electrofocusing. However, enzyme activity could be largely restored by the addition of the required zinc to the incubation medium.

Oxidation of cysteine and methione residues during IEF has been reported by Jacobs (1971) but can often be avoided by judicious use of antioxidants such as thiodiglycol or ascorbic acid. Park (1973) and Bunn (1973) ran dithionite through gel columns to prevent the oxygenation of deoxyhemoglobin during GEF.

Figure 10 shows a histochemical stain for the isoenzymes of lactate dehydrogenase (LDH) from chicken tissues after GEF. In this case, reduced pyrimidine nucleotides formed by the oxidation of lactate formed an insoluble product by reduction of tetrazolium salts.

Enzymes that act on large substrates such as proteins may be detected by zymogram techniques. In this method, the focused gel is overlaid with a high porosity gel, e.g., agarose, impregnated with a suitable substrate. This method is better suited to gel slabs than to cylinders. Vesterberg and Eriksson (1972) detected staphylokinase activity by overlaying focused gels with a layer of fibrinoclot containing plasminogen. The staphylokinase activity formed plasmin from the plasminogen to create a clear spot on the opaque fibrin plate. Similar *in situ* procedures may be used for autoradiographic detection of insoluble radioactive products. Further details of zymogram procedures may be found in review articles by Wadström and Smyth (1973).

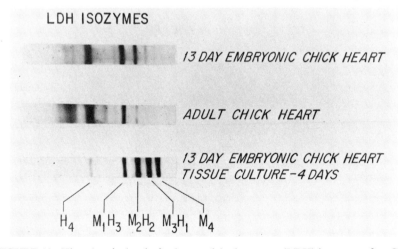

FIGURE 10. Histochemical stain for lactate dehydrogenase (LDH) isozymes after GEF in the pH 3–10 range. From Righetti and Drysdale (1973) by permission of the New York Academy of Sciences.

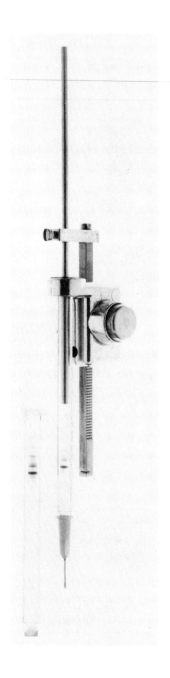

FIGURE 11. Fractionating device for gel rods. The gel tubes have a round collar at one end and a standard syringe taper tip at the other. After focusing, a syringe needle is attached to the tapered tip and the gel is extruded in uniform fractions through the needle with the aid of a Hamilton repeating dispenser. Note that no band distortion or compression occurs. From Bagshaw *et al.* (1973) by permission of the New York Academy of Sciences.

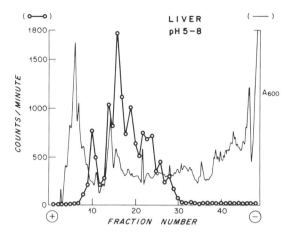

FIGURE 12. Analysis of mouse liver RNA polymerase by IEF. Solid line, scan at 600 nm of gel stained with Coomassie Blue; open circles, RNA polymerase assays of gel fractions. Anode is at the left. From Bagshaw *et al.* (1973) by permission of the New York Academy of Sciences.

C. Gel Fractionation

When no appropriate histochemical or autoradiographic method is available, enzymic or other biological activities can be detected in eluates from gel sections. A convenient and highly reproducible method for fractionating gels is illustrated in Figure 11. Gels are cast in syringelike tubes. After focusing, 10-μl portions of homogenized gel can be extruded through a No. 20 snub-nosed needle with the aid of a hamilton repeating dispenser.

Many enzymes can be assayed directly in the gel homogenate. Figure 12 shows a fractionation and assay of the isozymes of DNA-dependent RNA polymerase from mouse liver. Despite the large size of this enzyme (mol. wt. approximately 600,000) and the added DNA template (mol. wt. approximately 2×10^6), all reactants were apparently mutually accessible as evidenced by the high enzyme activities.

D. Recovery of Focused Zones

Generally, fairly good recoveries of focused proteins are obtained by squashing the gel and dialyzing against an appropriate buffer. The elution process may be accelerated by electrophoretic methods which also separate residual ampholytes from protein (Suzuki *et al.*, 1973). Ampholytes may also be separated from proteins by gel filtration (Brown and Green, 1970).

E. Markers

Colored markers are useful, not only for studying the development of the pH gradient, but also as indicators of equilibrium focusing. Bromophenol Blue and Fast Green, approximate pI 3.2, are useful for the most acidic pH ranges. Several other amphoteric dyes have also been suggested. These include Benzopurpurin, Congo Red Z, Chlorantine Fast Red, Hematein Carmine, Acid Fuchsian, Erie Garnet, and Orcein. Colored proteins of known pI are better indices of equilibrium focusing, but unfortunately only a few are readily available. Horse spleen ferritin (brown), pI 4.2–4.6 (Drysdale, 1974), is a useful marker protein and, because of its high molecular weight (440,000), is an excellent indicator of equilibrium focusing. Human or animal hemoglobins (p$I \sim 7$) (Drysdale et $al.$, 1971) and myoglobins (p$I \sim 7.5$) are ideal markers for middle pH ranges. Cytochrome c (horse heart) (p$I \sim 9.4$) is a useful indicator in basic pH ranges. By incorporating markers into the gel before polymerization the development of the pH gradient may be readily visualized. Equilibrium is attained when the anionic and cationic fronts coalesce.

V. APPLICATIONS

GEF is finding increasing use for qualitative and quantitative assessments of complex protein mixtures and for shedding new light on the structure of specific proteins. There are now several examples of proteins that appear homogeneous by other criteria but which have been resolved into multiple components by GEF [see Catsimpoolas (1973d)]. This section will describe a few such examples to demonstrate the resolving power of GEF and to indicate some of the many ways in which charge heterogeneity may arise in a protein.

A. Ferritins

My introduction to the improved resolution offered by isoelectric focusing came from studies of the iron-storage protein, ferritin. The protein moiety, apoferritin, had been extensively studied and at that time was thought to consist of a multimeric protein shell of 20 or 24 chemically identical subunits. Tissue-specific forms of apoferritins were known to exist, but they were thought to be homopolymers.

It was, therefore, with considerable surprise and some skepticism that I viewed the results of my colleague, Dr. Y. Niitsu, showing that crystalline preparations of this protein separated into multiple components when ex-

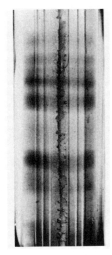

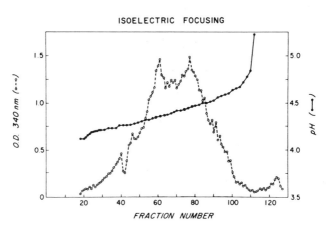

ISOELECTRIC FOCUSING

FIGURE 13. Fractionation of horse spleen ferritin by IEF in the pH range 4–6 in a sucrose density gradient. Courtesy of Dr. Y. Niitsu, Sapporo Medical School, Japan.

FIGURE 14. Elution of horse spleen isoferritins after IEF in the pH range 4–6 (see Figure 11). From Drysdale (1974) by permission of the *Biochemical Journal*.

amined by IEF. These components were clearly seen as brown bands by IEF in sucrose density gradients (Figure 13). Unfortunately, in attempting to isolate these components, we were frustrated by the loss of resolution from diffusion and mixing during the rather lengthy elution period (Figure 14). Because of this, and also the need for a speedier and less costly analytical procedure, we explored alternative systems and eventually developed a small-scale analytical procedure for IEF in polyacrylamide gel. This system gave highly reproducible pH gradients with LKB Ampholines. The gradients formed after about 2 hr but remained essentially stable for a further 18-hr period (Righetti and Drysdale, 1971). Such gradient stability was of paramount importance to ensure equilibrium focusing of the large ferritin molecule (mol. wt. 440,000) (Drysdale, 1974).

GEF helped to clarify the relationships of tissue-specific ferritins. As shown in Figure 15, most ferritins migrate as single bands on gel electrophoresis. The bands are, however, rather broad and have considerable overlap. When examined by GEF (Figure 16), all of these ferritins were resolved

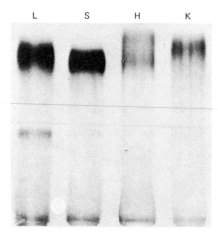

FIGURE 15. Analyses of ferritins from human liver (L), spleen (S), kidney (K), and heart (H) by electrophoresis in polyacrylamide gel at pH 8.6. From Powell *et al.* (1975) by permission of the *British Journal of Haematology.*

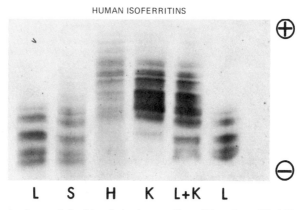

HUMAN ISOFERRITINS

L   S   H   K   L+K   L

FIGURE 16. Analyses of ferritins from human liver (L), spleen (S), kidney (K), and heart (H) by GEF in the pH range 4–6. From Drysdale *et al.* (1975) by permission of Elsevier Press.

into multiple components, with several common to most tissues. Those ferritins which were most difficult to resolve electrophoretically, e.g., liver and spleen, shared many common isoferritins. On the other hand, those which were readily distinguished by electrophoresis, e.g., liver and heart, shared only a few common isoferritins. These findings indicate that the different tissue ferritins are not homogeneous populations but families of isoferritins, many of which are common to several tissues. We have since found interesting shifts in isoferritin patterns in genetic disorders such as malignancy and iron storage disease (Alpert *et al.*, 1973; Powell *et al.*, 1974, 1975). Although the structural basis for the multiple forms in normal and diseased states has not yet been fully elucidated, present evidence indicates that the tissue isoferritins represent hybrid molecules consisting of different proportions of two or more dissimilar subunits (Drysdale, 1974; Drysdale *et al.*, 1975).

B. Hemoglobins

GEF has proved useful for the analysis of hemoglobin variants that are not readily distinguished by electrophoretic procedures. Figure 17 shows typical separations of some human hemoglobin variants. The good separation of hemoglobins A, F, and S is of particular interest for the diagnosis of sickle cell disease. By focusing in glass tubes, the relative amounts of the various forms may be rapidly assessed by densitometric scanning (Bunn, 1974). For rapid screening and comparative purposes, analyses by thin-slab electrofocusing may be preferred (Figure 7).

Another example of the remarkable resolving power of GEF is given from analyses of hemoglobins in different oxidation states. In preparations

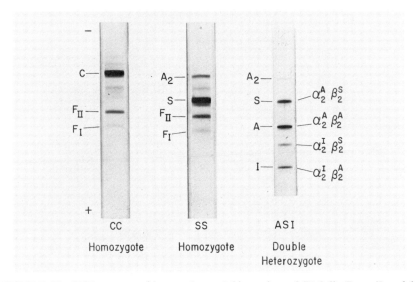

FIGURE 17. GEF patterns of human hemoglobin variants (pH 6–8). From Drysdale *et al.* (1971) by permission of ASP Biological Medical Press.

of partially oxidized hemoglobins, four forms, corresponding to oxyhemo-globin ($\alpha_2\beta_2$), methemoglobin ($\alpha_2{}^+\beta_2{}^+$), and intermediate half-oxidized forms $\alpha_2{}^+\beta_2$ and $\alpha_2\beta_2{}^+$, can be distinguished by GEF (Figure 18). Thus in the half-oxidized forms it is possible to detect the conformational difference arising from oxidation of the heme on either the α or β chain (Bunn and Drysdale, 1971; Park, 1973). Further evidence of the resolving power of GEF is given by the separation of deoxyhemoglobin and oxyhemoglobin (Bunn, 1973; Park, 1973). At equilibrium, deoxyhemoglobin was detected as a purple band with a higher pI than the red oxyhemoglobin. This is an elegant demonstration of the well-known Bohr effect.

It is interesting to note that while GEF will separate hemoglobins with rather small structural differences, e.g., hemoglobin Syracuse $\alpha_2\beta_2$ 143 His $\rightarrow$ Pro (H. F. Bunn, personal communication), it will not readily resolve hemoglobins A_2, C, or E, despite large differences in their primary structure. In such cases, analyses of separated subunits may be helpful. In reviewing the results given with hemoglobins it is significant that despite the added com-plexity often revealed by GEF, nearly all bands have been identified as different molecular structures or have been shown to have counterparts in electrophoretic or chromatographic procedures.

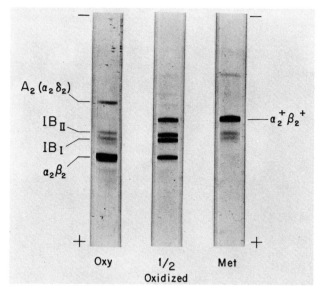

FIGURE 18. GEF patterns demonstrating partially oxidized hemoglobins IB$_I$ and IB$_{II}$, separated from oxyhemoglobin and methemoglobin (pH 6–8). From Bunn and Drysdale (1971) by permission of Elsevier Press.

C. Interacting Systems

Preliminary findings indicate that GEF may be of value in studying interacting protein systems or interaction with ligands. Hemoglobins again provide excellent models. For example, Park (1973) and Bunn (1973) have demonstrated subunit exchange and mixed tetramer formation from animal hemoglobins. These forms are not readily detected by other methods (Bunn, 1974). GEF has also been used to study the dimerization plane in deoxy- and oxyhemoglobin and its pH dependence.

GEF may also offer unique advantages in studies combining equilibrium and kinetic processes. For example, ligands or other reactants can be passed

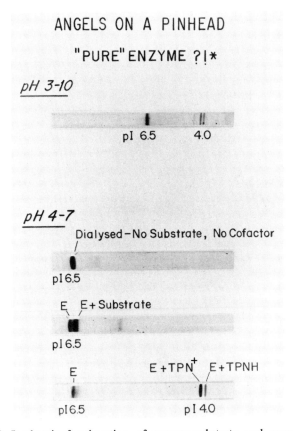

FIGURE 19. Isoelectric fractionation of enzyme–substrate and enzyme–coenzyme complexes (Drysdale and Littlefield, unpublished). From Righetti and Drysdale (1973) by permission of New York Academy of Sciences.

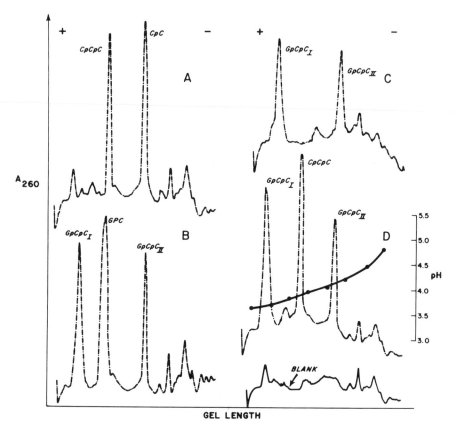

FIGURE 20. Isoelectric profiles of separation of GpC and CpC, GpCpC and CpCpC. Samples (0.15 O.D.) of each were mixed and cofocused in 5% polyacrylamide gels containing 2% ampholyte, pH 3–5. After an electrolysis period of 6 hr, the gels and a control containing only ampholytes were scanned at 260 nm. The pH gradient is indicated by the solid heavy line. From Drysdale and Righetti (1972) by permission of the American Chemical Society.

through a focused protein zone to investigate various kinetic parameters. Park (1973) used this approach to study ATP binding to hemoglobin and to calculate the stoichiometry of the complex.

GEF may also allow investigations of complexes of some enzymes with their substrates or cofactors. This possibility was suggested from studies of an apparently pure preparation of dihydrofolic acid reductase which unexpectedly resolved into three bands on GEF—the enzyme, enzyme–substrate, and enzyme–coenzyme complexes (Figure 19). The detection of such complexes will presumably depend on their stability in the pH range through which they move to their p*I*. In the above case, much of the heterogeneity in the enzyme preparation disappeared on prolonged electrolysis. Stinson (1974) has reported a similar time-dependent dissociation of a focused enzyme–ligand complex. In this case, he labeled free sulfhydryl groups on 3-phosphoglycerate kinase with a mercurial. The enzyme mercurial focused with a higher p*I* than the free enzyme but eventually was reduced to the free enzyme on prolonged electrolysis.

D. Nucleic Acids

Attempts to fractionate nucleic acis have given puzzling but slightly encouraging results. We have found that many oligonucleotides and polynucleotides form distinctive banding patterns when subjected to GEF in acidic pH ranges. Figure 20 shows banding patterns of some di- and trinucleotides. These simple nucleotides focused as single peaks. When rerun separately or together, each banded at the same p*I* as before. Measurements of apparent p*I* values indicated that these molecules were reaching a true

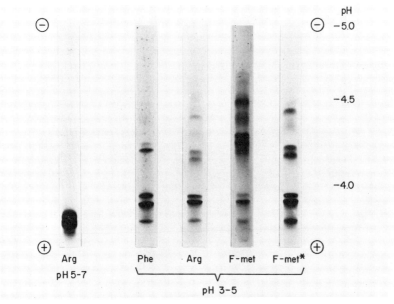

FIGURE 21. Fractionation of isoaccepting species of Phe, Arg, and F-Met tRNA by GEF. Samples (25 μg) were subjected to GEF for 6 hr in 5% polyacrylamide gels containing 2% ampholyte pH 3–5 or 5–7. In the case of F-Met*, the sample was incorporated directly into the gel before polymerization. In all others, the samples were applied to the top of the gel in a buffering layer of 2% ampholyte. From Drysdale and Righetti (1972) by permission of the American Chemical Society.

isoelectric state arising from protonation of ring nitrogens in A and/or C. Studies with highly purified tRNAs gave considerably more complicated results. Each of several highly purified isoaccepting species of tRNA resolved into multiple components in the pH range 3–6. Figure 21 shows banding patterns given by isoaccepting species of f-Met, Arg, and Phe tRNA from *Escherichia coli*. Subsequent analyses of isolated components indicated that these multiple forms were not degradation products, but molecules in different degrees of a reversible denaturation (Drysdale and Righetti, 1972).

More recently, we have investigated the possibility of separating discrete messenger RNA species by GEF. Preliminary results with alpha- and beta-globin messenger RNA from rabbit reticulocytes indicate that these RNA species could be separated from ribosomal and tRNA and from one another. However, as with tRNA, the basis of the fractionation is unclear and the possibility of interaction with ampholytes cannot be rejected (Drysdale and Shafritz, 1975).

E. Two-Dimensional Procedures

By combining GEF with other procedures it is often possible to obtain more information than from both techniques performed separately. For example, Dale and Latner (1969) combined GEF in the first dimension with gel electrophoresis in the second dimension to obtain protein maps that gave information on both the pI and the size of proteins in serum. In addition, this technique also allowed resolution of components that could not be

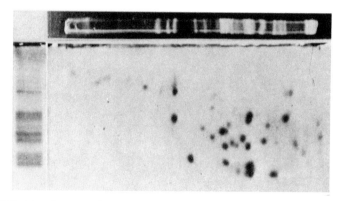

FIGURE 22. Protein map of wheat grain protein. Above is the pattern for GEF alone [pH 5 (left) to pH 9, 120-mm gel]. On the left is the pattern for starch gel electrophoresis alone (origin at top). Note that the two-dimensional technique resolves nearly twice the number of components separated by either method alone. From Wrigley (1970) by permission of the New York Academy of Sciences.

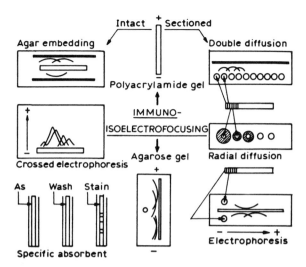

FIGURE 23. Various forms of immunoisoelectric focusing analysis in gels. From Catsimpoolas (1973c) by permission of New York Academy of Sciences.

clearly separated by either procedure. An example of a similar procedure for the analysis of grain protein is shown in Figure 22.

Kenrick and Margolis (1970) modified this technique by embedding the focused gel into the top of a concave 4.5–26% gradient slab of polyacrylamide. The electrophoresis is performed in the presence of sodium dodecyl sulfate.

Several methods have been developed for detecting isoelectrically focused proteins by immunological means. Figure 23 illustrates some of the procedures.

VI. DISCUSSION

The superior resolution given by conducting IEF in polyacrylamide gels rather than in sucrose density gradients is now widely recognized. Most of the early teething problems with the use of gels have now been satisfactorily overcome and several systems are now available for routine experimentation. The method is technically simple and is equal if not superior in resolution and sensitivity to other commonly used analytical procedures. Perhaps the greatest difficulty in the transition to polyacrylamide gel has been associated with pH-gradient instability. This problem has now largely been overcome with the development of systems in both cylinders and thin slabs in which pH gradients remain essentially constant for 18 hr after formation. Since most proteins will focus in these systems in less than 8 hr, one can now be fairly certain of obtaining genuine isoelectric fractionation long before the gradient begins to deteriorate. Ideally, it would be preferable to eliminate the problem altogether. As noted earlier, several factors may contribute to pH-gradient instability. One of the most important may be irregular conductivities of Ampholines in the pH range 5–8. It would, therefore, be most desirable to find additional ampholytes for IEF in this range. It would also be important to characterize carrier ampholytes more thoroughly. Although Ampholines have been available for almost 10 years, surprisingly little is yet known of their exact composition. The simplicity of the principle for their synthesis should facilitate investigations for other alternatives.

Many of the advantages of GEF are evident from some of the results presented here. Much of its appeal derives from it being an equilibrium method. Thus, the system is self-correcting and considerably less demanding in technique than many rate separations. Further, because of its built-in resolution, GEF has advantages over electrophoresis and ion-exchange chromatography where comparable separations often require specially tailored conditions that may not be equally effective for all components in the sample. In addition, the concentrating effects of GEF facilitates detection of trace components that may be too diffuse to detect by gel electrophoresis. For example, in red cell lysates it is possible to detect components that constitute as little as 0.2% of the total proteins (Drysdale et al., 1971).

Since the advent of GEF, there have been many demonstrations of heterogeneity in proteins by GEF that appear homogeneous by other criteria. Some of these differences may be rather subtle. In addition to differences in primary structure, other possibilities exist for charge heterogeneity. These include postsynthetic modifications, e.g., deamidation, ligand binding, differences in redox states in metalloproteins, and variations in prosthetic groups or nonprotein components. Considerable caution must therefore be exercised in interpreting banding patterns from GEF. However, the amount of new information to be gained should more than compensate for the trouble in separating the wheat from the chaff.

ACKNOWLEDGMENTS

I thank the many colleagues who made available both published and unpublished material and apologize to those whose work I have overlooked or have been unable to include.

I am particularly grateful to the following for sending me material for reproduction: H. F. Bunn, N. Catsimpoolas, H. Davies, C. Karlsson, K.

Keck, K. D. Rohnert, D. H. Leaback, A. J. MacGillivray, H. Rilbe. O. Vesterberg, and C. Wrigley.

This work was supported by grants CA 14406 and AM 16781 and 1775 from the National Institutes of Health.

VII. REFERENCES

Alpert, E., Coston, R. L., and Drysdale, J. W. (1973) *Nature* **242**: 194.

Awdeh, Z. L., Williamson, A. R., and Askonas, B. A. (1968) *Nature* **32**: 66.

Bagshaw, J. C., Drysdale, J. W., and Malt, R. A. (1973) *Ann. N.Y. Acad. Sci.* **209**: 363.

Blatter, D. P., Garner, F., van Slyke, K., and Brandlet, A. (1972) *J. Chromatogr.* **64**: 147.

Bours, J. (1973) *Sci. Tools* **20**: 29.

Brown, W. D. and Green, S. (1970) *Anal. Biochem.* **34**: 593.

Bunn, H. F. (1973) *Ann. N.Y. Acad. Sci.* **209**: 345.

Bunn, H. F. (1974) *Biochemistry* **13**: 988.

Bunn, H. F. and Drysdale, J. W. (1971) *Biochim. Biophys. Acta* **229**: 51.

Cannan, R. K. (1942). *Chem. Rev.* **30**: 295.

Catsimpoolas, N. (1971) *Anal. Biochem.* **44**: 427

Catsimpoolas, N. (1973) *Anal. Biochem.* **54**: 66.

Catsimpoolas, N. (1973b) *Sep. Sci.* **8**: 71.

Catsimpoolas, N. (1973c) *Ann. N.Y. Acad Sci.* **209**: 144.

Catsimpoolas, N., ed. (1973d) "Isoelectric focusing and isotachophoresis," *Ann. N.Y. Acad. Sci.* **209**.

Chrambach, A. and Rodbard, D. (1972) *Sep. Sci.* **7(6)**: 663.

Dale, G. and Latner, A. L. (1969) *Clin. Chem. Acta* **24**: 61.

Drysdale, J. W. (1974) *Biochem. J.* **141**: 627.

Drysdale, J. W. and Righetti, P. G. (1972) *Biochemistry* **11**: 4044.

Drysdale, J. W. and Shafritz, D. A. (1975) in *Advances in Isoelectric Focusing and Isotachophoresis* P. G. Righetti, ed. Elsevier Press, North Holland, Amsterdam.

Drysdale, J. W., Righetti, P. G., and Bunn, H. F. (1971) *Biochim. Biophys. Acta* **229**.

Drysdale, J. W., Hazard, J. T., and Righetti, P. G. (1975) in *Advances in Isoelectric Focusing and Isotachophoresis* P. G. Righetti, ed. Elsevier Press, North Holland, Amsterdam.

Felgenhauer, K. and Pak, S. J. (1973) *Ann. N.Y. Acad. Sci.* **209**: 147.

Finlayson, G. R. and Chrambach, A. (1971) *Anal. Biochem.* **40**: 292.

Gasparic, V. and Rosengreen, A. (1974) in *Isoelectric Focusing of Proteins and Related Substances* J. P. Arbuthnott and J. A. Beeley, eds. Butterworths, London.

Haglund, H. (1971) *Meth. Biochem. Anal.* **19**: 1.

Hayes, M. B. and Wellner, D. (1969) *J. Biol. Chem.* **244**: 6636.

Hjertén, S. (1962) *Arch. Biochem. Biophys. Suppl. 1* **98**: 147.

Jacobs, S. (1971) in: H. Peeters, (ed.) *Protides of the Biological Fluids* Vol. 18, Pergamon Press, New York, p. 499.

Kenrick, K. G. and Margolis, J. (1970) *Anal. Biochem.* **33**: 204.

Kolin, A. (1954) *J. Chem. Phys.* **22**: 1638.

Kolin, A. (1955) *Proc. Natl. Acad. Sci. U.S.A.* **41**: 101.

Kolin, A. (1958) *Meth. Biochem. Anal.* **6**: 259.

Latner, A. L., Parsons, M. W., and Skillen, A. W. (1970) *Biochem. J.* **118**: 299.
Leaback, D. H. and Rutter, A. C. (1968) *Biochem. Biophys. Res. Commun.* **32**: 447.
Malik, N. and Berrie, A. (1972) *Anal. Biochem.* **49**: 173.
Park, C. M. (1973) *Ann. N.Y. Acad. Sci.* **209**: 237.
Powell, L. W., Alpert, E., Isselbacher, K. J., and Drysdale, J. W. (1974) *Nature* **250**: 333.
Powell, L. W., Alpert, E., Isselbacher, K. J., and Drysdale, J. W. (1975) *Brit. J. Haematol.* **30**: 47.
Righetti, P. G. (1974) *Recent Advances in Isoelectric Focusing and Isotachophoresis*, Elsevier Press, North Holland, Amsterdam.
Righetti, P. G. and Drysdale, J. W. (1971) *Biochim. Biophys. Acta* **236**: 17.
Righetti, P. G. and Drysdale, J. W. (1973) *Ann. N.Y. Acad. Sci.* **209**: 163.
Righetti, P. G. and Drysdale, J. S. (1974) *J. Chromatog.* **98**: 271.
Rilbe, H. (1970) in (ed. H. Peeters) *Protides Biol. Fluids Proc. Colloq.* **17**: 369, Pergamon Press, New York.
Rilbe, H. (1973) *Ann. N.Y. Acad. Sci.* **209**: 11.
Riley, R. F. and Coleman, M. K. (1968) *J. Lab. Clin. Med.* **72**: 714.
Rodbard, D., Leviton, C., and Chrambach, A. (1972) *Sep. Sci.* **7**: 705.
Spencer, E. M. and King, T. P. (1971) *J. Biol. Chem.* **246**: 201.
Stinson, R. A. (1974) *Biochemistry* **13**: 4523
Suzuki, T. Benesch, R. E., Yung, S. and Benesch, R. (1973) *Anal. Biochem.* **55**: 249.
Svensson, H. (1961) *Acta Chem. Scand.* **15**: 325.
Svensson, H. (1962) *Acta Chem. Scand.* **16**: 456.
Udenfriend, S. Stein, S. Bohlen, P., Dairman, W., Leimgruben, W., and Weifele, M. (1972) *Science* **178**: 871.
Ui, N. (1973) *Ann. N.Y. Acad. Sci.* **209**: 198.
Vesterberg, O. (1969) *Acta Chem. Scand.* **23**: 2653.
Vesterberg, O. (1970) *Protides Biol. Fluids Proc. Colloq.* **17**.
Vesterberg, O. (1972) *Biochim. Biophys. Acta* **257**: 11.
Vesterberg, O. and Eriksson, R. (1972) *Biochim. Biophys. Acta* **285**: 393.
Vesterberg, O. and Svensson, H. (1966) *Acta Chem. Scand.* **20**: 820.
Vesterberg, O., Wadström, T., Vesterberg, K., Svensson, H., and Malmgren, B. (1967) *Biochim. Biophys. Acta* **133**: 435.
Vinogradov, S. N., Lowenkron, S., Andronian, M. R., and Bagshaw, J. (1973) *Biochem Biophys. Res. Commun.* **54**: 501.
Wadström, T. (1974) *Ann. N.Y. Acad. Sci.* **209**: 405.
Wadström, T. and Smyth C. J. (1973) *Sci. Tools* **20**: 7.
Wellner, D. (1971) *Anal. Chem.* **43**: 59A.
Williamson, A. R. (1973) in *Handbook of Experimental Immunology* (D. M. Weir, ed.) pp. 1–23, Blackwell, Oxford.
Wrigley, C. W. (1968) *J. Chromatog.* **36**: 362.
Wrigley, C. W. (1970) *Biochem. Genet.* **4**: 509.

PURIFICATION OF CHEMICALLY MODIFIED PROTEINS

5

ROBERT E. FEENEY AND DAVID T. OSUGA

I. INTRODUCTION

Protein chemists have long been interested in changing the chemical, physical, and biological properties of proteins by chemically altering their structure. In recent years, the application of modern knowledge of proteins and the uses of new chemical reagents and more sophisticated analytical techniques have made chemical modifications of proteins a major tool in protein chemistry (Hirs and Timasheff, 1972a; Means and Feeney, 1971). But, even with the availability of modern procedures, purification and analysis of the chemically modified products still remain as formidable tasks in using chemical modification in protein chemistry.

Probably the most important general rule to follow in the selection of the proper procedures for the purification and fractionation of modified proteins is that the selection must be on an operational basis. The method of purification must be selected with regard to the type of chemical modification, the manner in which modification is done, and the purpose of the modification. There are many choices of chemical-modification procedures for

ROBERT E. FEENEY AND DAVID T. OSUGA, Department of Food Science and Technology, University of California, Davis, California 95616.

achieving similar goals and the selections must be made with regard to the particular problem at hand. Since both the selection of the chemical-modification procedure and the procedures for purification of the product are based on related operational concepts, any plans for chemical modification must include consideration of the requirements for purification of the products.

II. GENERAL TYPES AND PURPOSES OF CHEMICAL MODIFICATION

Proteins are chemically modified for many different reasons and the same methods may be used in different ways for different purposes. Consequently, although classifications into different types of modifications can be based on their chemistries (Means and Feeney, 1971), it is more appropriate in this chapter to classify the modifications on the basis of their purposes. We have divided modifications into the following broad general groups: modification of groups essential for changing biological properties, changing physical properties, blocking deteriorative reactions, introduction of labels, and allowing reversible modifications. Classifications of this nature are, of course, never absolute and there are no sharp lines between the different types.

A. Modification to Change Biological Activity

Modifications of protein side chains have been a powerful tool in providing information concerning the chemical basis for enzyme functions. The direct method of X-ray crystallography provides the absolute data, but chemical modification probably remains the easiest and relatively inexpensive way to identify "essential groups" and their functions in solution. The chemical-modification approach has also been applied to other types of biologically functional proteins as well as to those with enzymatic activity. There are many different ways in which the biological activities of proteins can be changed by chemical modification. Changes in biological function can be attributed frequently to modification of active-site groups, but it is necessary to screen carefully the many other possible reasons for such changes. For example, in the case of enzymes, there are many instances in which modification causes large changes in K_m or pH optima without much effect on the capability of the groups in the catalytic site to function. Changes in K_m or of $\overline{V}$ in an enzyme are not always easy to interpret.

Other changes in biological activity which can be obtained by chemical modification include increases in immunogenicity or introduction of a specific hapten, inactivation of toxins with retention of immunogenicity (i.e., production of toxoids), increases or decreases in susceptibilities to hydrolysis by proteolytic enzymes, and increases or decreases in transport properties (e.g., cell adsorption).

Specificity of the chemical modification is obviously important for most modifications used to determine "essential groups," but the control of the extent of modification may frequently be more critical. One or two side-chain groups of a particular type, such as sulfhydryl or amino groups, may be much more reactive than others of the same type. In some cases, the procedure of modification might be controlled so as to modify these particular groups and almost none of the others of the same type. In other instances, several of a larger number of the same type of group may be modified much faster or more slowly than the others. But, even when a larger number of groups are modified at a faster or slower rate than others, a kinetic analysis of the changes caused by the modification may give valuable information. Such a differential modification of the fast-reacting amino groups of turkey ovomucoid was employed in our laboratory to calculate the number of amino groups essential for its trypsin-inhibitory activity (Figure 1) (Haynes et al., 1967). When such modified samples are fractionated, heterogeneity can be expected due to differences in the number of amino groups modified. These problems will be discussed further in Section III.C.

The techniques that are used to react more specifically with and label the active centers or catalytic sites of enzymes and other biologically active proteins are particularly important to enzyme chemists (Means and Feeney, 1971). Specific reactions in reactive centers are usually accomplished by utilizing the protein's affinity for a reagent that resembles the protein's own substrate or effector group. Such affinity labeling reagents may be considered to have two separate parts. One part has a high affinity for the reactive center of the protein. The other part is a group capable of forming a covalent bond with a group in the vicinity of the active center. A photoaffinity labeling reagent also has the affinity part in its molecule, but the reactive part is generated by photoactivation after the reagent has combined noncovalently with the enzyme. What has been named a K_{cat} enzyme inhibitor (Rando, 1974) is an affinity labeling reagent whose reactive part is generated during the specific catalysis of the enzyme. The photoaffinity labeling might be expected to cause unusual problems in purification and analysis because photoactivation produces free radicals capable of forming derivatives not usually encountered on protein modification.

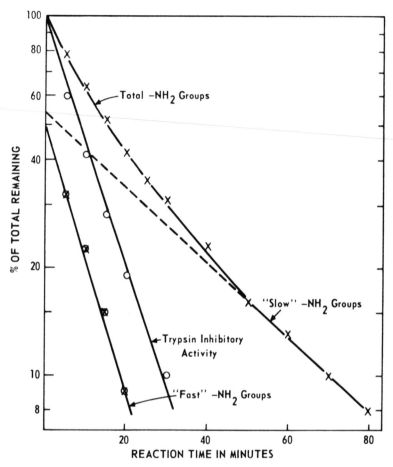

FIGURE 1. Semilogarithmic plot for loss of amino groups and loss of trypsin-inhibitory activity in turkey ovomucoid by modification with trinitrobenzenesulfonic acid. From Haynes *et al.* (1967).

B. Modification to Change Physical Properties

A frequent purpose of chemical modification is to change the physical properties of the protein. This is usually done by forming products that change the net charge on the protein, either to make it more cationic or more anionic, by forming intra- or intermolecular cross-links, or by scission of naturally occurring disulfide cross-links.

Changes in the charges on a protein usually have rather large effects on its solubility and in many cases will dissociate noncovalently linked polymers. The most commonly used procedures are those which involve reaction with the ε-amino group of lysines so as to make the protein more acidic, i.e., to make the isoelectric point more acidic. Acetylation, for example, changes an amino group into a neutral amide. Substitution of the benzene ring of tyrosine by iodination or nitration lowers the pK of the tyrosine hydroxyl and also is useful for making a protein more acidic. More basic proteins can also be caused by esterification of carboxyl groups, but esterification is not commonly used for this purpose.

Bifunctional reagents with two reactive groups are used to introduce both inter- and intramolecular cross-links into proteins (Fasold *et al.*, 1971). They are frequently used to stabilize tertiary structure, to prepare cross-linked models, to measure molecular distances between interacting groups, to estimate numbers of different size polymeric forms of monomers (Figure 2), or to determine the physical proximities of proteins in physical structures such as membranes. Each of these types would require special methods if the products were to be purified. With soluble, nonpolymeric proteins the products may be primarily intramolecular or intermolecular (to form polymers) depending upon the conditions used.

One of the easiest ways to change extensively the physical properties of a protein is by reduction of its disulfide bonds. Only the disulfide bonds of proteins are susceptible to mild reduction. When the newly exposed sulfhydryl groups are blocked by alkylation, the reformation of disulfide groups can be prevented. Other changes of properties may also be obtained by varying the nature of the alkylating reagent or by using oxidation to form a new disulfide bond with a sulfhydryl-containing reagent. Other types of products may be obtained by using different reagents, such as sulfite, in scission of the disulfide bonds. Fractionation of most of these products is usually done by methods utilizing the changes in charge resulting from the modification. Changes in solubilities may, of course, be both advantageous and disadvantageous.

C. Modifications to Block Deteriorative Reactions

Modifications are also done to block certain functional side chains so as to prevent deteriorative effects during purification, handling, or storage. One example is the blocking of the ε-amino groups of lysine to prevent the carbonyl–amine reaction with carbonyl compounds, in particular with sugars, causing the deteriorative reaction known as the Maillard reaction or nonenzymatic browning (Feeney *et al.*, 1975). Another example is the blocking of sulfhydryl groups to prevent their oxidation by the formation of intra- or intermolecular disulfide bonds or by carboxymethylation. Products of these types are not usually purified.

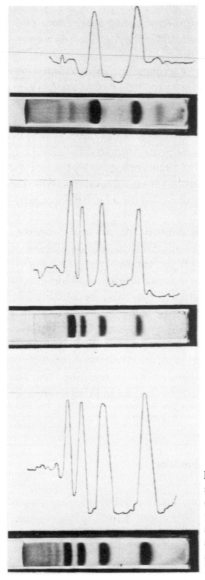

FIGURE 2. Patterns of sodium dodecyl sulfate-polyacrylamide gel electrophoresis of dimethyl suberimidate cross-linked oligomeric proteins stained by Coomassie brilliant blue and their densitometer tracings: (bottom) glyceraldehyde-3-phosphate dehydrogenase, (middle) aldolase, (top) tryptophan synthetase B. From Davies and Stark (1970).

D. Introduction of a Label

Still other modifications are done to introduce some type of labeling in a protein. This label can be used as a signal to monitor changes in the protein's structure under various conditions (reporter groups), simply to identify it when mixed with other proteins, to locate an active site, or as a reference target (isomorphic replacement) in X-ray crystallography. Common labels used as reporter groups are those with chromophores or with fluorescent groups, and the ones for X-ray studies are heavy-metal atoms.

Radioactively labeled reagents are frequently used to label proteins for many purposes. When a radioative label is used as a tracer of the protein, little or no change in physical properties is usually desired. Reductive methylation of amino groups with ^{14}C-labeled formaldehyde appears especially well suited for this purpose (Means and Feeney, 1968; Rice and Means, 1971). The substituent methyl groups are comparatively small and small differences in charge due to methylation of the molecule are usually unimportant. In contrast, labeling by iodination of tyrosines causes large changes in charge which can cause the individual protein molecules with different amounts of iodination (and consequently different charges) to separate during purification.

E. Reversible Modifications

Another general classification of modifying reagents include those which are termed reversible reagents because the modifying group can be removed either by changing the environment of the protein, such as changing pH, or by the use of a second chemical treatment of the protein. These are particularly important in those many instances where the purpose of the modification is to reversibly mask one type of group in a protein from reaction with another reagent used to modify the other groups. For example, a first reagent reversibly modifies sulfhydryl groups and a second reagent modifies amino groups. If the sulfhydryls had not been protected, they would have been irreversibly modified by the second reagent. Finally, the first reagent would be removed so that the sulfhydryl groups are regenerated.

The reductive splitting of disulfide bonds in proteins is often a reversible reaction. By simply exposing a solution of the reduced protein to air, the original native structure of the protein may often be attained again.

The use of reversible modification to separate oligomers of different types of subunits will be discussed in Section VI.A.

There are, of course, many variations of the above types of modification as well as many other types which are used to varying extents by protein chemists.

III. SOME PROBLEMS ENCOUNTERED IN CHEMICAL MODIFICATION

Problems encountered in chemical modification may be minor or major ones, depending upon the type of chemical fractionation to be employed and the conditions to which the product will be exposed. Some of the problems can be frequently avoided, but there are a number of problems that make purification methods impossible or useless. Consequently, it is sometimes more desirable to use the product without any attempt at purification. A general appreciation of the problems will materially assist in preventing unnecessary difficulties.

A. Denaturation

The occurrence of denaturation on chemical modification of proteins is always a serious possibility. In some instances, however, it is immaterial whether or not the product is denatured, and perhaps it may even be desirable to have it so. These particular instances will not be discussed here.

1. Denaturation Caused by the Modification Procedure

When the conditions or reagents used to modify proteins are harsh, denaturation may accompany the modification. Extremes of pH, for example may be the most suitable way to maintain a reagent or amino acid side chain in a reactive state, but this may have an undesirable effect on protein structure. It may be necessary to use agents which are commonly classified as deforming or denaturing agents, such as urea, for unfolding the protein in order to increase the accessibility and, hence, the reactivity of the various protein side chains. When removal of the agents results in the refolding of the protein to its original structure, this may be no problem.

In some cases, organic solvents may be required to help maintain a reagent in solution. In a few cases the solvent may serve as a reactant and must be present in high concentration in order to influence favorably the equilibrium, such as in the case of esterification of proteins in anhydrous MeOH/HCl.

Proteins, of course, vary widely in their relative resistances to different kinds of denaturing conditions. Many enzymes and soluble globular proteins are sufficiently labile that extensive modification is not possible with many methods. Fortunately, extensive modification is often unnecessary, and incompletely modified products that can be considered native are still useful. The reason this is possible with many of these proteins with enzymatic activities is that the groups involved in their catalytic function are on, or near the surface of, the molecule. Hence, it is unnecessary to unfold the molecule to modify the side chains directly concerned. When it is necessary to open up the protein in order to modify residues situated internally, such as disulfide bonds, drastic changes may occur in the structure.

2. Liability of the Product

One of the most important things to remember when chemically modifying a protein is that the product is now a new chemical substance. As such, it may or may not have properties similar to the original starting material. This is particularly important when considering the possibility of obtaining a denatured product. Even though the conditions employed for chemical modification might not irreversibly denature the protein, the product may be labile to many conditions to which the original native protein was relatively stable. The product may, however, be more stable to certain conditions than the original native protein. Changes in lability as a result of the modification must, therefore, be considered in the purification of the chemically modified protein. If possible, lability studies should be made on the modified product before instituting chemical and physical fractionation procedures. This would be particularly the case if the original native protein was relatively stable at moderately acidic or alkaline pHs and the modified product was now to be exposed to these conditions.

It is therefore necessary to choose the correct reagent for the modification which will not provide too harsh conditions for the initial process and which will not provide a product which is too labile and therefore cannot be handled or used. For example, egg white lysozyme can be easily esterified with MeOH/HCl and the esterified product can be easily crystallized. Most proteins, however, will not take this procedure and esterifications must be made by procedures which can be done near neutrality.

B. Reversibility of Modification

Modification reactions which are readily reversible under certain conditions are highly desirable procedures for many purposes in protein chemistry. On the other hand, they may also cause considerable problems if the purification of the product requires that it be handled under conditions where reversibility of the reaction readily occurs. With certain reactions that are readily reversible, it may be difficult or even impossible to purify the modified product and it may be necessary to use it with only the simplest and crudest methods of purification under mild conditions. An example of such a relatively reversible product is an esterified protein. Ester groups will slowly hydrolyze at only slightly alkaline pH values.

C. Heterogeneity of Products

Heterogeneity of the chemically modified product is probably one of the greatest difficulties facing the chemical-modification expert. It may not be possible to obtain homogeneous products or at best to obtain only relatively homogeneous products. Although many heterogeneous products may retain the identifiable biological function of the original native protein, kinetic constants for these functions of such a heterogeneous population may be difficult to interpret. Sometimes the goal must be merely to exclude certain types of heterogeneity rather than to obtain a "homogeneous" preparation. Four types of heterogeneity are commonly observed.

1. Heterogeneity due to Substitution of Different Numbers of the Same Type of Side Chain

A common type of heterogeneity is due to incomplete modification of a particular desired type of side chain, such as amino groups. The modification may be deliberately incomplete. In such instances, the product may have a population distribution varying from completely unmodified molecules to molecules with all their amino groups modified. Usually, however, the distribution is much narrower and it might be expected that if, for example, 50% of the total amino groups were modified, individual molecules might mainly vary from 30% to 60% modification (Figure 3). The distribution in a population will vary with the protein as well as with the modification procedure. Some groups are modified much faster than other groups and, indeed, as described in Section II.C, this relative reactivity has been used as a means for determining the number of essential groups (Ray and Koshland, 1962; Haynes et al., 1967) (Figure 1).

The heterogeneity might also include populations with similar numbers of modified amino groups but modified on different residues of the peptide chain. Separations of these different types of heterogeneous populations would usually be incomplete.

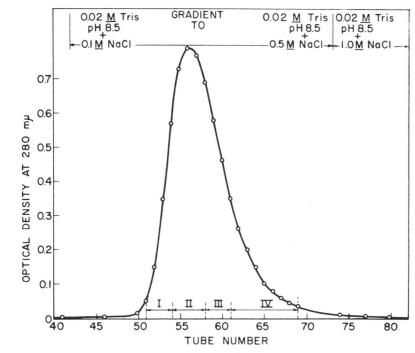

FIGURE 3. DEAE-cellulose chromatography of partially acetylated turkey ovomucoid. The starting buffer was 0.02 M tris, pH 8.5. From Stevens and Feeney (1963).

2. Heterogeneity due to the Modification of Different Types of Side Chains

Heterogeneity may also result from nonspecificity of the chemical-modification procedure. Most chemical modifications of side chains involve attack by electrophiles or oxidation. Specificity is, therefore, usually only relative and dependent upon the conditions. Frequently, the nonspecificity may be sufficiently small that there is no serious problem or the nonspecific modifications may result in products which can be easily separated from the other desired modified products. Therefore, the knowledge of the non-specificities of the particular methods employed must be carefully considered when the purification procedures are selected. For example, when tetranitro-methane is used to nitrate tyrosines, cross-linking between aromatic groups may occur, presumably due to free-radical formation. Some of the cross-links could produce polymers that might be separated from the other molecules by gel filtration. It is evident that in some instances purification of the protein may prove to be a major project in its own right!

3. Heterogeneity due to Partial Denaturations

As discussed above, the original native protein or the product may be denatured. Only part of the product may be affected, and it may be necessary to remove denatured forms by some type of purification procedure. The particular properties of the material may be used to separate it from these denatured forms, such as the use of affinity chromatography for a modified enzyme which still retains its activity. Usually, insolubility will separate much of the denatured material. However, in a few instances, the denatured forms may persist even after extensive fractionations.

4. Heterogeneity from Side Reactions of a Chemical Nature

The conditions for modification may be such that side reactions which are essentially chemical in nature may occur. These will be discussed below in Section III.E.

D. A Product with Properties Very Similar to Those of the Original Native Protein

A frequently desired objective in chemical modification is to obtain a product that has properties very similar to those of the original native material. This is particularly true when the objective is to differentiate between modifications that block a particular type of side chain so that it cannot function and modifications that may also change the charge or hydrophobicity to a marked extent. It is also true when the objective of the modification is to introduce a radioactive label into the protein without changing its properties significantly. When the ε-amino groups of lysines are alkylated by reductive dimethylation, a chemically modified product is obtained with physical properties very similar to those of the original protein (Means and Feeney, 1968). It is nearly impossible to fractionate these products by most methods, although more recently available techniques involving hydrophobic chromatography and isoelectric focusing at pH 9–10 might have some success. Fractionation of such products would not usually be attempted.

E. Side Reactions of a Chemical Nature Occurring During Modification

Under the heading of chemical side reactions can be included the very many reactions which involve changes in the primary structure of the protein as a consequence of the extreme conditions or as a result of the modification

TABLE I
Possible Chemical Side Reactions on Protein Modification

Groups	Treatment	Effects
Peptide bonds	Alkaline pH	Hydrolysis
	Acidic pH	N → O Acyl shift
Thiol groups	Oxidation	—S—S—, Acids
Disulfide bonds	Reduction	—SH, Mispairing
	Alkaline pH	Hydrolysis, β-elimination
Methionyl groups	Oxidation	Oxy sulfurs
Amide groups	Alkaline pH	Hydrolysis
O-Glycosyl	Alkaline pH	β-Elimination
O-Phosphate	Alkaline pH	β-Elimination

(i.e., this group does not include the nonspecific reactions mentioned in Section III.C, above). These can cause very critical problems and can introduce serious difficulties in purification of the modified product. However, judicious selection of reagents and strict control of reaction conditions can usually avoid most of these side reactions.

Table I lists the main possible side reactions encountered. The most serious problems are encountered when alkaline pHs are employed. A pH as high as 10 is usually detrimental and should be avoided. For some proteins, of course, this pH will also cause denaturation. Denaturation may be more easily seen than some of the side reactions.

Many of the chemical side reactions give products which will change the charge on the molecule. These include the hydrolysis of amide groups to carboxyl groups and β-elimination of various groups (phosphates, sugars, and disulfides). The β-eliminated product can react with other groups producing such compounds as lanthionine and lysinoalanine (Bohak, 1964).

IV. ANALYSIS OF CHEMICALLY MODIFIED PROTEINS

One of the first subjects a synthetic organic chemist considers before he starts a synthesis is the availability of methods for analysis of the desired products. Unfortunately, this is not as easy to do in the chemical derivatization of proteins because it is usually more difficult to predict the properties of the product. The problem is sometimes complicated by the necessity of either maintaining the protein in its "native" state for analysis or by chemical reactions occurring during or prior to analysis of the protein by such means as hydrolysis to amino acids. Nevertheless, analysis of the chemically modified product is one of the most important aspects of the chemical modification of proteins. In addition to the avoidance of the problems outlined in Section III, there are quite a number of analytical procedures that should be considered before a protein is chemically modified.

A. Analysis of Amino Acid Residues

1. Direct Analysis of the Modified Protein

The direct analysis of the chemically modified protein without acid or basic hydrolysis should be used with all chemically modified proteins when applicable. A spectrophotometric analysis for detection of changes in the uv region should be done, especially because tyrosine, tryptophan, and disulfides (which contribute appreciably to uv spectra) are groups that are often intentionally modified. These groups are also frequently unintentionally modified because of their reactivities and the nonspecificity of the techniques. The chemical modification may also introduce new absorbing groups, and it is usually critical that those be detected and quantitated. Several chemical analytical procedures that do not necessitate acidic or basic hydrolysis are available for certain of the amino acid side chains (Means and Feeney, 1971).

2. Amino Acid Analyses on Acid or Base Hydrolyses

Amino acid analyses on acid or base hydrolysates should always be determined regardless of whether or not the desired products are stable to these conditions, unless it is known that these conditions will produce other subsequent side reactions which would obscure the results. In some instances, like acetylation, this is usually useless because the desired products, and, nonspecifically made products, are unstable to acidic or basic hydrolysis.

Even nonspecificity, therefore, would not be seen. In many other instances the loss of the product can be easily seen after acidic hydrolysis. Such is the case with the modification of arginines with 1,2-cyclohexanedione (Liu *et al.*, 1968). The arginine analysis on acid hydrolysates is quite accurate. Proteins usually do not contain a large number of arginines so that on a percentage basis the loss of a few can easily be seen.

3. Detection of Loss of an Amino Acid with the Formation of a New Product

This procedure of determination of the loss of a particular amino acid and the formation of a new product is obviously one of the better ways of quantitating chemical modifications. An example is seen in Figure 4 and

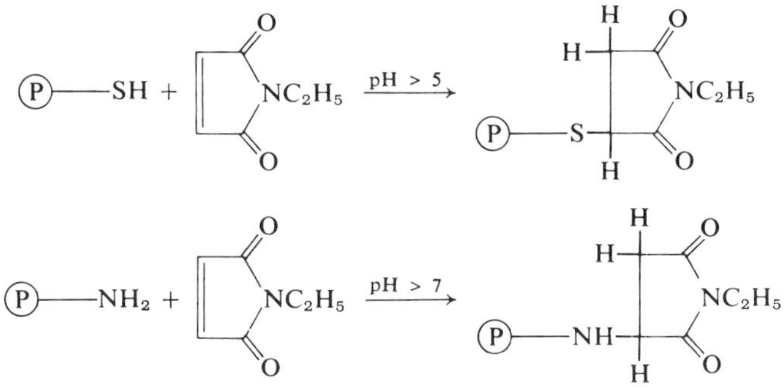

FIGURE 4. Reaction of protein, P, sulfhydryl group (top) and amino group (bottom) with N-ethylmaleimide. From Means and Feeney (1971).

Table II for the modification of proteins with N-ethylmaleimide. In proteins lacking sulfhydryl groups, the modification is reasonably specific for the ε-amino groups of lysine. It can be seen in Table II that as lysine disappears there is a new product of lysine [ε-N-(2-succinyl)lysine] as well as ethylamine. The ethylamine is produced on acid hydrolysis from the N-ethylsuccinimidyl derivative of lysine.

Products that are not amino acids have been determined by special methods or by further modifications to the original or a different amino acid. For example, in alkaline β-elimination (Section III.E), glycosyl groups from O-seryl linkages in glycoproteins form 2-aminopropenic acid residues (Downs and Pigman, 1969). The glycoproteins are hydrolytically cleaved at these residues in weak acid to form pyruvyl residues at what would have been the NH_2-terminal end at the point of cleavage (Figure 5). The pyruvyl residue can now be converted to alanine. O-Glycosyl linkages with threonines,

TABLE II
Products of Ribonuclease A after Treatment with
N-Ethylmaleimide[a]

	Residues (moles/mole protein)			
	Control		pH 8	pH 7	pH 6
Lysine	9.7	(10)[b]	2.2	4.6	7.1
Histidine	3.7	(4)[b]	3.4	3.2	3.8
Ethylamine	—		6.2	3.8	2.5
ε-N-(2-Succinyl)lysine	—		6.4	4.0	2.3

[a] From Brewer and Riehm (1967).
[b] Theoretical values.

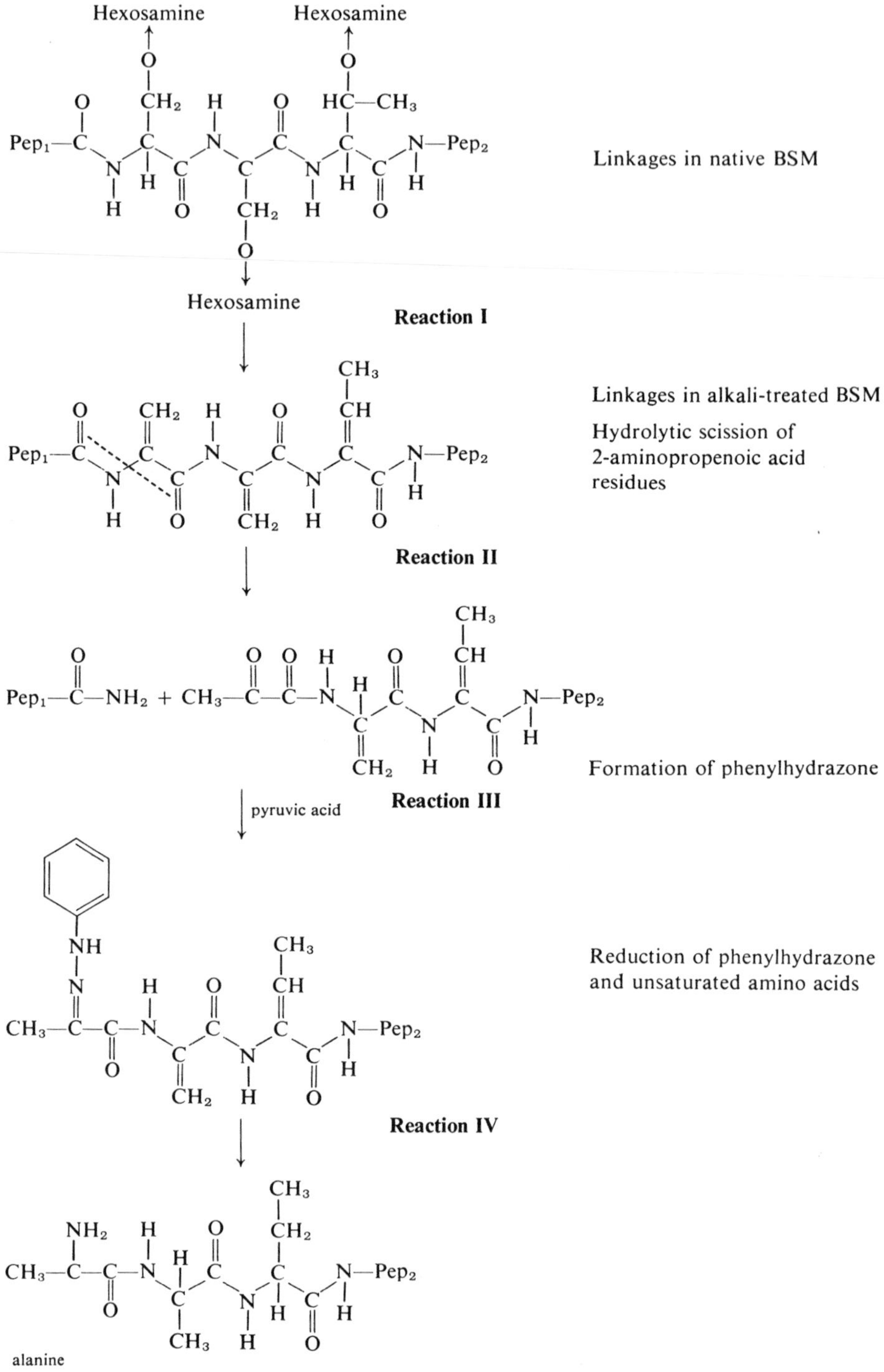
Hexosamine
Hexosamine
Hexosamine
Linkages in native BSM
Reaction I
Linkages in alkali-treated BSM
Hydrolytic scission of 2-aminopropenoic acid residues
Reaction II
Formation of phenylhydrazone
pyruvic acid
Reaction III
Reduction of phenylhydrazone and unsaturated amino acids
Reaction IV
alanine

O-phosphoryl linkages with threonines, O-acyl derivatives with serine (e.g., tosylchymotrypsin), or disulfides can also undergo β-elimination in alkali to form unsaturated amino acids (Ahmed *et al.*, 1973; Ako *et al.*, 1974; Cavallini *et al.*, 1970). With the antifreeze glycoproteins from Antarctic fish the unsaturated residues can be determined by their absorption at 241 nm because there is only a small absorbance at this wavelength in the native glycoprotein (Ahmed *et al.*, 1973). The pyruvic acid formed by hydrolytic scission in acid has also been determined enzymatically (Ako *et al.*, 1974).

In other instances simpler products may be detected, although the separation of products may either be only partial or may require special chromatographic procedures. The 3-N-methyl derivative of histidine has been separated from histidine by conventional amino acid analysis. The reaction with methyl *p*-nitrobenzenesulfonate methylates a ring nitrogen of one (the active site) of three histidines in α-chymotrypsin by an affinity labeling procedure (Nakagawa and Bender, 1970; Ryan and Feeney, 1975). The histidine peak is reduced to approximately two with a hump on the side which corresponds to approximately one (0.9) group (Ryan and Feeney, unpublished data). This hump is the correct position for a methylhistidine. In another type of modification, the products are well separated only by altering the elution scheme for chromatography of the amino acids. Figure 6 shows the separation of the dimethyl and the isopropyl derivatives of the ε-amino groups of lysine on the conventional amino acid analytical columns with slight changes in the buffer system to further differentiate the peaks of isopropyllysine, dimethyllysine, and lysine.

A slightly different approach to chemically modified products is seen in the case of the oxidized products of methionine (Figure 7). Methionine can be S-alkylated with iodoacetic acid to the carboxymethylsulfonium derivative. It can be oxidized with hydrogen peroxide to the sulfoxide or with performic acid to the sulfone. The sulfoxide is converted to methionine again on acid hydrolysis, but the sulfone is stable to acid hydrolysis. The sulfone can be measured by amino acid analysis and used as a measure of the sulfoxide

---

FIGURE 5. The cleavage of bovine submaxillary mucin (BSM) into glycopeptides. In Reaction I, native bovine submaxillary mucin was treated with 0.2 N NaOH for 4 hr at 45°C. The carbohydrate side chains were removed by the β-elimination reaction with alkali leaving the unsaturated amino acids in the protein core (β-E-bovine submaxillary mucin). In Reaction II, β-E-bovine submaxillary mucin was hydrolyzed at pH 2.2 for 1 hr at 100°C. In Reaction III, a fraction from a G-25 Sephadex column of the product of Reaction II was treated with phenylhydrazine hydrochloride for 48 hr at 4°C. Reaction IV is the reduction of the phenylhydrazone using PdCl$_2$ and NaBH$_4$. Threonine is represented as having a β-eliminated form that is resistant to acid. Actually, the differences between threonine and serine are not this large. Adapted from Downs and Pigman (1969).

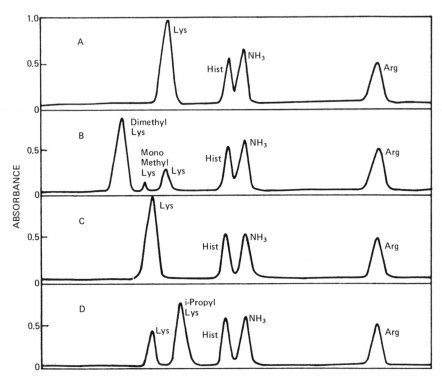

FIGURE 6. Chromatographic patterns of reductively alkylated ribonuclease (RNase) on a Technicon 21-hr autoamino acid analysis system with modified elution buffers. A and C are unmodified RNase; B and D are methylated and isopropylated RNase, respectively. From Means and Feeney (1968).

(plus sulfone) present at the time of carboxymethylation. In these treatments, iodoacetic acid does not affect sulfoxide residues, and the carboxymethyl-sulfonium derivative, produced by the action of the iodoacetic acid on unoxidized methionine, is resistant to performic acid. The carboxymethyl-sulfonium derivative is unstable to the acidic hydrolytic condition, giving rise to some of the original methionine, S-carboxymethylhomocysteine, and homoserine lactone. The summation of these products can be used to determine the amounts of the various products, depending upon which type of modification is initially used. For example, partial oxidation of the methionine to the sulfoxide can be determined by alkylation of the remaining methionines; alternatively, partial alkylation can be measured by oxidation of the unmodified methionines with performic to form the sulfone (Feinstein and Feeney, 1967).

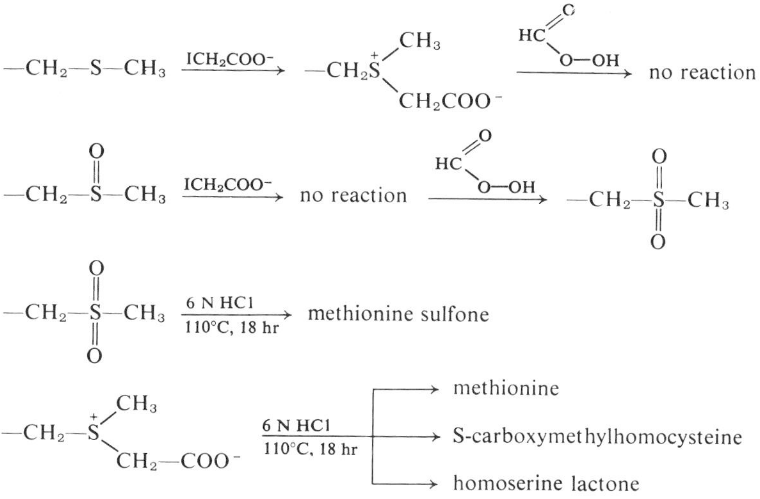

FIGURE 7. Reaction of methionine and methionine sulfoxide with iodoacetic acid and performic acid, respectively, and their hydrolysis products. From Means and Feeney (1971).

4. Introduction of Special Chromogens

When groups are introduced that have a special absorption maximum easily differentiated from other protein groups, this can, of course, be used for the direct determination of the number of groups introduced. There are many examples of these and no review is necessary at this point.

5. The Use of Radioisotopically Labeled Reagents

The use of isotopically labeled reagents is an invaluable tool in synthetic organic chemistry and, similarly, it is also useful in the chemical modification of proteins because the introduction of the radioactive label can be used in a variety of ways, the simplest of which is the direct determination of the amount of radioactivity introduced. This can also be compared with the degree of modification estimated by other methods as a check on specificity (Section II.D).

6. Other Instrumental Methods

There are now many instrumental methods available for study and analysis of the physical and chemical structures of proteins (Hirs and

Timasheff, 1972b, 1973). The use of these would most likely depend upon their availability in a particular laboratory, the desired degree of sophistication, and the financial support for the investigation. For example, mass spectrometry, separately, or with gas–liquid chromatography, will probably be used more in measuring protein derivatization; these procedures are now commonly available.

V. TYPICAL EXAMPLES OF THE PURIFICATION OF CHEMICALLY MODIFIED PROTEINS

A. Heterogeneity Commonly Encountered in the Introduction of Groups Causing Changes in Charges

The chromatographic profile of an acetylated preparation of turkey ovomucoid shows the wide spreading of the chromatographic peak typical of this modification technique (Figure 3). This wide peak was correlated with the inactivation of the ovomucoid in its biochemical action of combining with and inhibiting bovine trypsin (Stevens and Feeney, 1963). The extent of modification under these conditions was at that time estimated by means of the electrophoretic patterns and observations of charge changes to be approximately 40–50%. This would be approximately six groups, or half of the total of 12 amino groups in turkey ovomucoid. Later studies using the kinetic approach to determine the number of essential amino groups (Figure 1) (Haynes et al., 1967) showed that there was only one amino group required for the inhibitory activity and that this was one of the faster-reacting groups. Separation of those species of molecules that had the active-site amino group modified would have been difficult at that time. Other inhibitors, modified by enzymatic cleavage of the peptide bond of the specific combining groups, have since been separated electrophoretically (Laskowski and Sealock, 1971). This specific enzymatic modification, not possible chemically, has now been achieved with turkey ovomucoid (Osuga et al., 1974). The use of a reversibly modified intermediate is described in Section VI.D.

Isoelectric focusing should be applicable to the fractionation and purification of chemically modified proteins that differ only by small numbers of charges introduced. Unfortunately, this method would probably not differentiate between the modifications of similar types of residues at different positions in the peptide chain. For example, when a monoiodinated derivative of adrenocorticotropic hormone was separated into one fraction by isoelectric focusing, hydrolysis with chymotrypsin and separation of the peptides

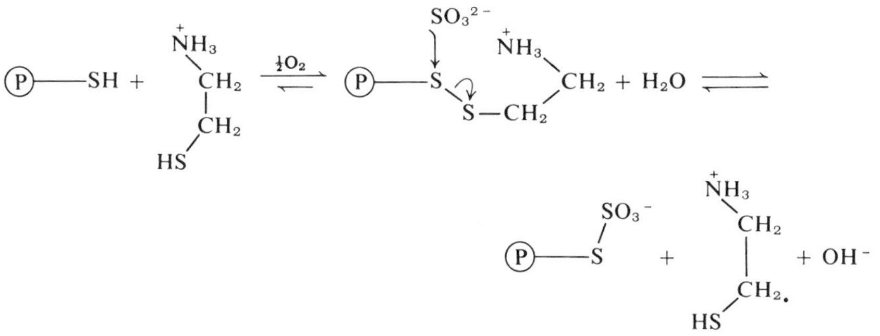

FIGURE 8. S-Sulfonation of a protein sulfhydryl to produce a charged product. β-Mercaptoethylamine is required to catalyze the reaction. From Means and Feeney (1971).

showed that the preparation was a mixture of two different products, each monoiodinated on a different single tyrosine (Rae and Schimmer, 1974).

Separations of species generated by other chemical modifications have been more successful than the above described one for acetylation. When there are only a few particularly reactive groups in the protein molecule, less heterogeneity is usually encountered. With such groups as sulfhydryls there may be one or, at the most, two very reactive sulfhydryls. In these cases, the modification of a sulfhydryl with a reagent that introduces a charge change, and in which the reaction is highly specific for that protein, can be used to separate the modified protein from the unmodified ones. Alkylation with iodoacetate or S-sulfonation (Figure 8) are such reactions.

Cytochrome c (horse heart) is one of the proteins that can be obtained in its oxidized and reduced states, two interconvertible forms differing by only a single charge. Dixon and Thompson (1968) separated the oxidized and reduced forms on CM-cellulose and discussed the relative contributions of the change in charge and a change in conformation to chromatographic separation.

B. Modifications with No Change in Charge

When the modification does not result in a significant change in charge (Section III.D), purification may be difficult. If there is a large change in hydrophobicity, molecular shape, or size, however, these properties can be used advantageously. One such example is in the fractionation of the mercury dimers of serum albumin (Figure 9). These can be separated on basis of solubility.

$$\text{P}\!-\!\!-\!\!\text{SH} + Hg^{2+} \rightleftharpoons \text{P}\!-\!\!-\!\!\text{SHg}^+ + H^+$$

$$\text{P}\!-\!\!-\!\!\text{SHg}^+ + \text{P}\!-\!\!-\!\!\text{SH} \rightleftharpoons \text{P}\!-\!\!-\!\!\text{S}\!-\!Hg\!-\!S\!-\!\!-\!\!\text{P} + H^+$$

FIGURE 9. Formation of mercaptalbumin mercuric dimer. The protein sulfhydryl reacts with mercury which then reacts with a second protein sulfhydryl to form the dimer. From Means and Feeney (1971).

C. Some Special Procedures for Chromatography

1. Affinity Columns

The specific adsorption characteristics of one protein to another, or between a protein and a low-molecular-weight substance, can be used for protein purification (Cuatrecasas and Anfinsen, 1971; Feinstein, 1971; Means and Feeney, 1971; Wilchek and Givol, 1973). Affinity chromatography is the name that has been given to the technique which exploits such specific and reversible interactions for chromatographic purification of proteins as well as other substances.

Our laboratory has used affinity chromatography for the purifications of riboflavin-binding protein from chicken egg white (Blankenhorn et al., 1975) and of protein inhibitors and proteolytic enzymes that form highly associated specific complexes with one another (Means et al., 1974; Ryan and Feeney, 1975). In the purification of the riboflavin-binding protein, the protein is chromatographed on a column containing a support to which a derivative of the flavin is covalently attached. In the purification of the protein inhibitors and proteolytic enzymes, our method is an adaptation of that of Feinstein (1970) for the chromatography of bovine α-chymotrypsin on a column containing covalently linked turkey ovomucoid, an inhibitor of the α-chymotrypsin. Chromatography on columns containing turkey ovomucoid is done with native α-chymotrypsin and two chemically modified derivatives, 3-methylhistidine-57-chymotrypsin and anhydrochymotrypsin, which have little or no catalytic activities but retain their capacities to form complexes with the turkey ovomucoid (Figure 10). (The synthesis and purification of the methylchymotrypsin are further described in Section VI.C). Chromatography of the turkey ovomucoid is also possible on preparations of the methylchymotrypsin attached to an insoluble support (Ryan and Feeney, 1975). The use of this chemical derivative of α-chymotrypsin for the chromatography is superior to the use of α-chymotrypsin because the latter causes peptide bond hydrolysis, while the inert derivative does not.

Affinity chromatography has been extensively used with columns prepared with antibodies to a particular protein or to an introduced chemical group. When an antibody is to be made against the low-molecular-weight

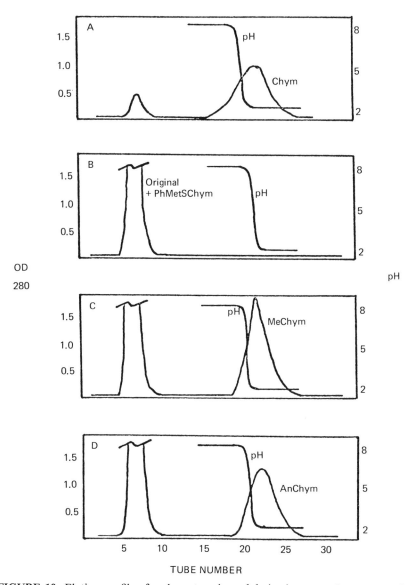

FIGURE 10. Elution profile of α-chymotrypsin and derivatives on turkey ovomucoid–Sepharose 4B column. Optical density and pH plotted vs. tube number. (A) Native chymotrypsin (Chym), (B) phenylmethanesulfonyl chymotrypsin (PhMetSChym), (C) crude methylchymotrypsin (MeChym), and (D) crude anhydrochymotrypsin (AnChym). The column was 2.0 × 23 cm; initial and wash buffer was pH 8.0; 0.05 M tris with 0.2 M NaCl and 0.03 M CaCl$_2$; eluting of sample was done with 0.03 M CaCl$_2$, 0.2 M NaCl in 0.1 M formic acid. Adapted from Ryan and Feeney (1975).

substance introduced into the proteins, a primary requirement is that the group which is introduced into the protein must have the structures prerequisite for good immunogenicity. This usually requires some type of aromaticity, although other structures such as certain carbohydrate side chains may also be employed. This approach can be particularly useful in purifying some chemically modified proteins because it may be possible to prepare antibody columns which are highly specific for the group introduced into the protein by the modification.

Affinity chromatography with antibody columns has been reported to be especially useful for the specific isolation of peptides containing modified residues from selectively modified proteins (Wilchek *et al.*, 1971; Wilchek and Givol, 1973). In this method the chemically modified protein is hydrolyzed into peptides, usually by proteolysis, and the modified peptides are separated from the entire peptide mixture by a one-step isolation on a column containing antibodies to the group introduced by the modification. Two examples are the isolation of arsanilazotyrosyl-containing peptides from arsanilazo-*N*-succinylcarboxypeptidase A by antiazobenzenearsonate antibodies and the isolation of a DNP-lysyl-containing peptide from mono-DNP bovine pancreatic ribonuclease by anti-DNP antibodies (Figure 11).

More recently, columns with attached lectins (Lis and Sharon, 1973) have been extensively used for both the analysis and the purification of glycoproteins. In some instances these might be useful for proteins with carbohydrate groups introduced by the modification. Our laboratory has used a lectin-inhibition assay in studying the β-elimination of carbohydrate

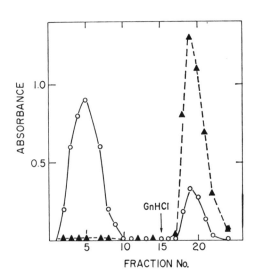

FIGURE 11. Isolation of a single peptide from a performic acid-oxidized DNP-RNase digested by trypsin on an anti-DNP-Sepharose column. Absorbance at 280 nm (○) and 360 nm (▲) plotted against fraction number. DNP-peptide eluted by the 6 M guanidine·HCl (GnHCl). From Wilchek *et al.* (1971).

from the antifreeze glycoprotein from Antarctic fish blood serum (Ahmed *et al.*, 1973).

2. *Affinity Elution*

Affinity elution has been the name used for the elution of a protein from an ion exchanger by alteration of the charge of the protein through the introduction into the buffer of a ligand that specifically combines with the protein. Illingworth (1972) purified yeast isocitrate dehydrogenase by adding AMP to the eluting buffer. Although affinity elution does not appear to have had any application to chemically modified proteins, it might be useful in cases where the chemical modification would cause an appreciable difference in the affinity for the ligand.

3. *Hydrophobic Chromatography*

Proteins bind hydrocarbons to varying extents, depending upon the structure of the hydrocarbon and the structure of the protein (Mohammad-zadeh-K. *et al.*, 1969). Bovine serum albumin and bovine β-lactoglobulin, for example, bind over 100 times the amount of heptane as chicken egg white lysozyme binds under the same conditions. The binding of heptane by lysozyme, however, increases severalfold in acidic solutions and nearly 100-fold upon reduction of the disulfides and alkylation of the resulting sulfhydryl groups. Hydrocarbon binding would appear, in general, a useful analytical method to determine changes in protein conformation.

Shaltiel (1973) and coworkers (Er-el *et al.*, 1972) have developed an homologous series of hydrocarbon-coated agaroses to provide an adjustable tool for the chromatographic separation of proteins. Figure 12 shows the effects of hydrocarbon chain lengths on the chromatography of glycogen phosphorylase *b*. Phosphorylase *a* was retained on Sepharose with a short hydrophobic side chain while phosphorylase *b* was retained by a longer side chain (Shaltiel, 1973). Shaltiel has suggested that hydrophobic chromatography should be useful for separating regulatory proteins which are designed to assume more than one stable conformation.

VI. THE USE OF CHEMICAL MODIFICATION AS A TOOL FOR PURIFICATION OF PROTEINS

The material which has been discussed so far in this chapter has concerned the purification and analysis of chemically modified proteins. Chemical

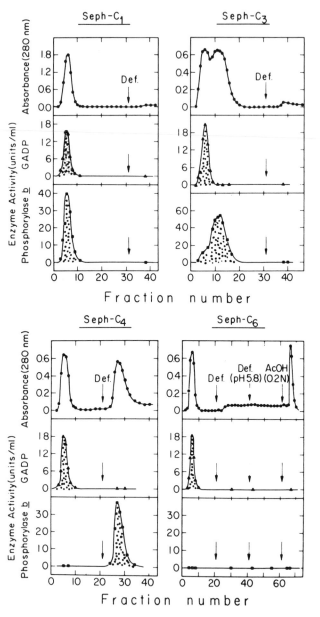

FIGURE 12. Separation of phosphorylase b from glyceraldehyde-3-phosphate dehydro-genase (GADP) by hydrophobic chromatography. Columns were Sephadex (Seph) with various hydrocarbons attached, i.e., Seph-C_1 = Seph-NH-CH_3; Seph-C_3 = Seph-NH-$CH_2CH_2CH_3$. Def. and the arrow indicates the initiation of a buffer change. From Er-el *et al.* (1972).

modification is also used as a tool for the purification of proteins. In some cases a combination of modification for the sake of modification and for the purification of modified proteins is also done. One of the important requirements in the use of chemical modification for this purpose is usually that the modification must be reversible in order to obtain again the protein in purified form after fractionation. However, in some instances, the reversibility is not required when modification is used primarily as an analytical procedure or when it is used to modify only part of a population of protein molecules.

A. Purification by Reversible-Complex Formation

When a reversible complex of a protein with some substance has properties which can be used to separate proteins by conventional methods, the formation of such complexes can be used to purify and perhaps to study the proteins. Borate forms readily reversible complexes with many carbohydrates and gives the complex a negative charge which can be used to electrophorese the complexes (Schwimmer *et al.*, 1956).

Our laboratory has applied the formation of such borate complexes to the separation of a group of closely related glycoproteins from fish blood serum. These glycoproteins were the antifreeze glycoproteins from Antarctic fishes (Feeney, 1974). They have no charged amino acid side chains and differ from one another mainly only by molecular size.

These glycoproteins form borate complexes which can be separated by electrophoresis in acrylamide gels (Figure 13). After electrophoresis, the glycoproteins can be eluted from the gels and the borate removed by dialysis

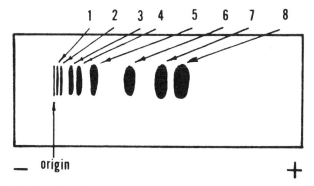

FIGURE 13. Polyacrylamide electrophoresis of Antarctic fish blood glycoproteins. Discontinuous-buffer system at pH 8.6 was used: the gel buffer was tris-citrate and the bridge buffer was borate. Adapted from Vandenheede (1972).

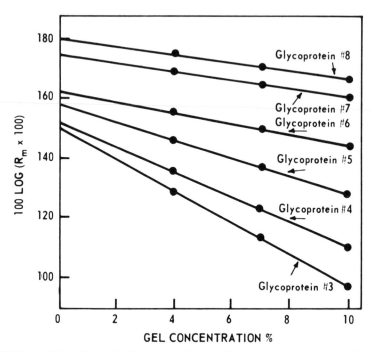

FIGURE 14. The effect of gel concentration during electrophoresis using the borate discontinuous-buffer system on the relative migration (*Rm*) of the glycoproteins isolated from Antarctic fish blood. From DeVries *et al.* (1970).

or by passing through a gel exclusion agent. Purified proteins are obtained which differ primarily on the basis of molecular weight. The smaller proteins electrophorese more rapidly than the larger ones under these conditions. The purified proteins could also have their molecular weights rather accurately estimated by their relative migrations on electrophoresis in gels containing different concentrations of acrylamide and gel cross-linking agents (Figure 14). Thus, the capacity to complex borate can be used to purify the proteins and also to estimate their relative sizes.

B. Reversible Chemical Modification for Separating Hybrids of Variants of Oligomeric Proteins

Gibbons *et al.* (1974) recently described a method for separating hybrids of chromatographically indistinguishable variants of oligomeric forms of

TABLE III
Use of Reversible "Chromatographic Handles" to Separate Hybrids
of Native and Inactivated Subunits[a]

1. Separation of catalytic (C_N) and regulatory (R) subunits of C_2R_3 (C is a trimer, R is a dimer; hybrid has 12 polypeptides)
2. Inactivation of C_N by reductive alkylation with pyridoxal phosphate $\rightarrow C_P$
3. Acylation of C_P with 3,4,5,5-tetrahydrophthalic anhydride $(T) \rightarrow C_{PT}$
4. Formation of hybrids with C_N, C_{PT}, and $R \rightarrow C_{PT}C_{PT}R$, $C_NC_{PT}R$, C_NC_NR
5. Chromatography to separate hybrids on basis of charge
6. Removal of T from hybrids by incubation at pH 6

[a] Adapted from Gibbons et al. (1974).

aspartate transcarbamylase (Table III) (Figure 15). When hybrids formed between native and mutant or specifically modified proteins are studied, it is often difficult to obtain them in pure form. These workers have been able to separate these hybrids by reversibly introducing charges by the reaction of amino groups of one variant with tetrahydrophthalic anhydride followed by ion-exchange chromatography. The acidic substituent groups can be removed under the mild conditions of pH 6 for 24 hr and the desired hybrids are then obtained. This procedure gives complete restoration of enzyme activity following the removal of the blocking groups. By the use of this procedure, they were able to show that cooperativity exists in the enzymelike molecules containing active chains in only one subunit.

C. The Use of Chemical Modification to Separate Products with Different Biochemical Activities

The separation of chemically modified proteins on the basis of different specific reactivities of modified products has been utilized in our laboratory for the purification of bovine α-chymotrypsin which has had its histidine-57 methylated on the nitrogen of the 3 position. This was possible by the use of a "double-fractionation" procedure. Methylchymotrypsin was prepared by the methylation of the active-site histidine of chymotrypsin with methyl p-nitrobenzenesulfonate (Nakagawa and Bender, 1970; Ryan and Feeney, 1975) (Figure 16). In this modification, the substrate serves as an affinity label and the reactive-site histidine-57 was methylated on a nitrogen. The methylchymotrypsin was shown to have properties sufficiently similar to native chymotrypsin that it would form a strongly associated complex with turkey ovomucoid, an inhibitor of chymotrypsin (Section V.C.1). The

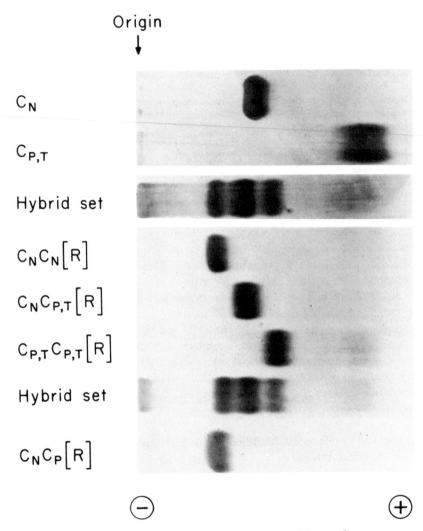

FIGURE 15. Electrophoretic patterns of intersubunit hybrid sets of aspartate trans-carbamylase. C_N, a hybrid containing one native catalytic subunit; C_{PT}, pyridoxylated and tetrahydrophthaloylated subunit; samples 3 and 7 are a hybrid set. The original sample which was fractionated on DEAE-Sephadex; samples 4, 5, and 6 are the fractions; the bottom is the product of deacylation of sample 5. From Gibbons *et al.* (1974).

particular adaptation of the method used in our laboratory was to use only the amount of the substrate reagent as needed to obtain approximately 50% of the product. This partial modification was done to limit the amount of reagent so as to give the highest degree of specificity for the reactive-site

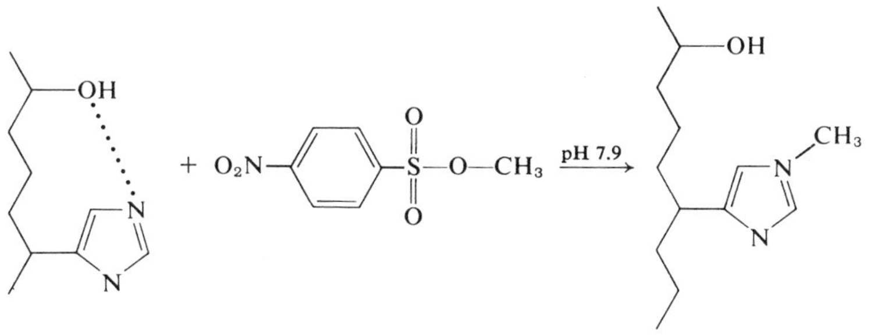

FIGURE 16. Methylation of histidine in α-chymotrypsin by methyl p-nitrobenzene sulfonate. From Ryan and Feeney (1974).

histidine. The preparation which now contained 50% native chymotrypsin and 50% methylchymotrypsin was then chemically modified with tosyl-chloride in slight excess. This modified the active-site serine in the native chymotrypsin. Serine of the N-methylchymotrypsin was not modified because the catalytic activity of the enzyme in the methylchymotrypsin is extremely low. The mixture now contained 50% of tosylchymotrypsin and 50% of the methylchymotrypsin. When this was passed through a turkey ovomucoid–Sepharose column, the methylchymotrypsin was adsorbed while the tosylchymotrypsin passed through the column. The methylchymotrypsin could then be regenerated from the column by lowering the pH (Figure 10).

D. Reversible Modification to Allow for Another Modification

A reversible chemical modification of a protein may sometimes be used to achieve a second chemical or enzymatic modification and to help in the purification of the products. A good example of such a series of reactions is the reduction of disulfide bonds in order to obtain enzymatic hydrolysis and resynthesis of reactive-site peptide bonds of the human basic pancreatic trypsin (Kunitz) inhibitor (Tschesche et al., 1974). When many proteinase inhibitors are treated with the proteinases, one particular peptide bond of the inhibitor is reversibly hydrolyzed. This peptide bond is considered to be a "reactive-site bond" at the combining site of the inhibitor for the enzyme (Means et al., 1974). The Kunitz inhibitor, however, is not hydrolyzed.

Figure 17 shows the series of reactions used to accomplish the hydrolysis of the reactive-site peptide bond in the Kunitz inhibitor. The single disulfide bond reduced under these conditions is the key to the equilibrium of the hydrolysis of the peptide bond. By scission of this disulfide bond, it is pos-

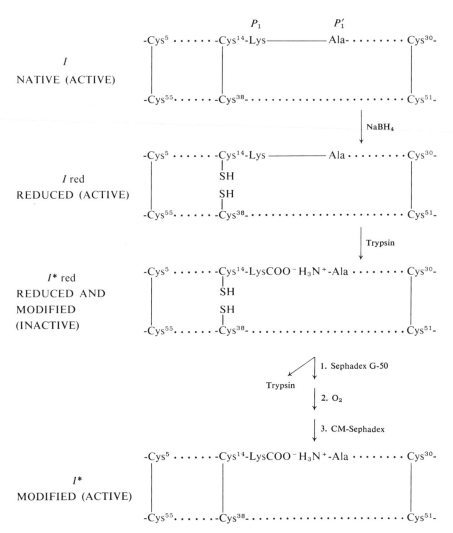

FIGURE 17. Chemical and enzymatic reactions in preparation of semisynthetic Lys
^{15}-Ala^{16}-cleaved Trypsin-Kallikrein Inhibitor (Kunitz). *I* is native inhibitor; *I** is inhibitor with reactive-site peptide bond hydolyzed. P_1 and P'_1 are the two amino acids on
the combining site. Adapted from Tschesche *et al.* (1974).

sible to hydrolyze the peptide bond and fractionate the products. When the
disulfides of the purified product with both the disulfide bond and peptide
bond broken are allowed to reoxidize, this heretofore unattainable intermediate product with the peptide bond split now has its peptide bond
resynthesized when incubated with enzyme.

E. The Use of Cross-Linking Agents to Determine Degree of Polymerization and Localization of Proteins in Tissues

A unique application of the cross-linking of proteins was used by Davies and Stark (1970) for the determination of the extent and kinds of polymers present in a polymeric protein. Dimethyl suberimidate was used to modify several oligomeric proteins. After modification, the proteins were examined by electrophoresis in polyacrylamide containing sodium dodecyl sulfate, which causes proteins to move at rates proportional to their molecular sizes. When oligomers were composed of identical subunits, the number of principle species observed was identical to the number of subunits, but when the method was applied to two proteins composed of dissimilar subunits there were different numbers of polymers obtained (Figure 2).

ACKNOWLEDGMENTS

The researches providing background for this article were supported by grants from the NIH (HD-00122-10 and AM-13686-06), and from the FDA (FD-00568-01). The authors are particularly grateful to Drs. H. B. F. Dixon and Gary E. Means for their critical suggestions and to Ms. C. A. Howland for her detailed technical editorial assistance.

VII. REFERENCES

Ahmed, A. I., Osuga, D. T., and Feeney, R. E. (1973) *J. Biol. Chem.* **248**: 8524.
Ako, H., Foster, R. J., and Ryan, C. A. (1974) *Biochemistry* **13**: 132.
Blankenhorn, G., Osuga, D. T., Lee, H. S., and Feeney, R. E. (1975) *Biochim. Biophys. Acta*: in press.
Bohak, Z. (1964) *J. Biol. Chem.* **239**: 2878.
Brewer, C. F. and Riehm, J. P. (1967) *Anal. Biochem.* **18**: 248.
Cavallini, D., Federici, G., Barboni, E., and Marcucci, M. (1970) *FEBS Lett.* **10**: 125.
Cuatrecasas, P. and Anfinsen, C. B. (1971) *Annu. Rev. Biochem.* **40**: 259.
Davies, G. E. and Stark, G. R. (1970) *Proc. Natl. Acad. Sci. U.S.A* **66**: 651.
DeVries, A. L., Komatsu, S. K., and Feeney, R. E. (1970) *J. Biol. Chem.* **245**: 2901.
Dixon, H. B. F. and Thompson, C. M. (1968) *Biochem. J.* **107**: 427.
Downs, F. and Pigman, W. (1969) *Biochemistry* **8**: 1760.
Er-el, Z., Zaidenzaig, Y., and Shaltiel, S. (1972) *Biochem. Biophys. Res. Commun.* **49**: 383.
Fasold, H., Klappenberger, J., Meyer, C., and Remold, H. (1971) *Angew. Chem.* **10**: 795.
Feeney, R. E. (1974) *Am. Sci.* **62**: 712.

Feeney, R. E., Blankenhorn, G., and Dixon, H. B. F. (1975) *Adv. Protein Chem.* **29**: 135.

Feinstein, G. (1970) *FEBS Lett.* **7**: 353.

Feinstein, G. (1971) *Naturwissenschaften* **58**: 389.

Feinstein, G. and Feeney, R. E. (1967) *Biochim. Biophys. Acta* **140**: 55.

Gibbons, I., Yang, Y. R. and Schachman, H. K. (1974) *Proc. Natl. Acad. Sci. U.S.A.* **71**: 4452.

Haynes, R., Osuga, D. T. and Feeney, R. E. (1967) *Biochemistry* **6**: 541.

Hirs, C. H. and Timasheff, S. N. (1972a) *Meth. Enzymol.* **25B**: 1.

Hirs, C. H. and Timasheff, S. N. (1972b) *Meth. Enzymol.* **26C**: 1.

Hirs, C. H. and Timasheff, S. N. (1973) *Meth. Enzymol.* **27D**: 1.

Illingworth, J. A. (1972) *Biochem. J.* **129**: 1119.

Laskowski, M. and Sealock, R. W. (1971) in *Enzymes 3rd ed.* 3 (P. D. Boyer, ed) pp. 375–473, Academic Press, New York.

Lis, H. and Sharon, N. (1973) *Annu. Rev. Biochem.* **42**: 541.

Liu, W. H., Feinstein, G., Osuga, D. T., Haynes, R., and Feeney, R. E. (1968) *Biochemistry* **7**: 2886.

Means, G. E. and Feeney, R. E. (1968) *Biochemistry* **7**: 2192.

Means, G. E. and Feeney, R. E. (1971) *Chemical Modification of Proteins*, Holden-Day, San Francisco.

Means, G. E., Ryan, D. S., and Feeney, R. E. (1974) *Acc. Chem. Res.* **7**: 315.

Mohammadzadeh-K., A., Smith, L. M., and Feeney, R. E. (1969) *Biochim. Biophys. Acta* **194**: 256.

Nakagawa, Y. and Bender, M. L. (1970) *Biochemistry* **9**: 259.

Osuga, D. T., Bigler, J. C., Uy, R. L., Sjoberg, L., and Feeney, R. E. (1974) *Comp. Biochem. Physiol.* **48B**: 519.

Rae, P. A. and Schimmer, B. P. (1974) *J. Biol. Chem.* **249**: 5649.

Rando, R. (1974) *Science* **185**: 320.

Ray, W. J., Jr. and Koshland, D. E., Jr. (1962) *J. Biol. Chem.* **237**: 2493.

Rice, R. H. and Means, G. E. (1971) *J. Biol.Chem.* **246**: 831.

Ryan, D. S. and Feeney, R. E. (1974) *Bayer Symp.* **V**: 378.

Ryan, D. S. and Feeney, R. E. (1975) *J. Biol. Chem.* **250**: 843.

Schwimmer, S., Bevenue, A., and Weston, W. J. (1956) *Arch. Biochem. Biophys.* **60**: 279.

Shaltiel, S. (1973) in *Metabolic Interconversion of Enzymes* (E. H. Fischer, E. H. Krebs, H. Neurath, and E. R. Stadtman, eds.) pp. 379–392, Springer-Verlag, Berlin.

Stevens, F. C. and Feeney, R. E. (1963) *Biochemistry* **2**: 1346.

Tschesche, H., Jering, H., Schorp, G., and Dietl, T. (1974) *Bayer Symp.* **V**: 362.

Vandenheede, J. R. (1972) *Primary Structure and Mechanism of Action of a Freezing-Point Depressing Glycoprotein from Antarctic Fish.* Ph.D. Thesis, University of California, Davis, Calif.

Wilchek, M. and Givol, D. (1973) in *Peptides* (H. Nesvaba, ed.) pp. 203–222, North-Holland, Amsterdam.

Wilchek, M., Bocchini, V., Becker, M., and Givol, D. (1971) *Biochemistry* **10**: 2828.

CHROMATOGRAPHIC   6
PEAK SHAPE
ANALYSIS*

ELI GRUSHKA

I. INTRODUCTION

The shape of chromatographic peak is of utmost importance since it can inform the scientist of the processes which occur in the column where the separation is taking place. Consequently, the shape can yield a wealth of information dealing with such diverse topics as the recognition of strongly overlapped peaks and adsorption–desorption rate constants. It is, therefore, surprising that the majority of the papers dealing with chromatographic theory discuss mainly retention characteristics and/or zone broadening. The peak shape is fundamentally more important since not only can the retention time and the peak width be obtained from it, but the validity of the usual

---

* *Editor's Note*: This article on "Chromatographic Peak Shape Analysis" by Eli Grushka does not contain material dealing directly with the separation of proteins. Nevertheless peak shape analysis as described in this work can provide important information to the protein chemist especially in matters pertaining to resolution of strongly overlapping peaks in electrophoresis, chromatography, and sedimentation. In addition, the peak shape carries physicochemical information related to the particular type of separation employed.

ELI GRUSHKA, Department of Chemistry, State University of New York at Buffalo, Buffalo, New York 14214.

assumption made in chromatography (namely, Gaussian shape; infinitely fast mass transfer through the stationary–mobile phase interface) can be ascertained by analyzing the shape.

Early research concentrated on proving that chromatographic peaks can be approximated by the Gaussian shape. (For a general review see Giddings (1965), Chapter 2.) This is particularly true of the plate model (at the limit of large plate number) developed by Martin and co-workers (Martin and Synge, 1941; James and Martin, 1952) and by Gleuckouf (1955). The much more powerful mass balance approach also attempted to introduce Gaussian peaks (namely, Lapidus and Amundson, 1952; van Deemter et al., 1956). The stochastic approach [McQuarrie (1963) and citation in Giddings (1965)] showed that the peak shape is not always Gaussian or symmetric, although in the limit of long time, the Gaussian shape is again obtained.

In the early 1960s many scientists recognized that an exact expression of the peak shape might be difficult to obtain, and research was directed toward extracting some parameters which would, at the same time, be indicative of the zone shape and be related to experimental conditions. Bocke and Parke (1962) showed that the statistical moments of the concentration profile of the peak can be found from the mass-balance equations which describe the chromatographic system. Statistical moments are the desired parameters since they can be related to the peak shape and, most importantly, they can show how the shape depends on the experimental setup. To date, several other groups have used moment analysis to discuss zone-shape characteristics in various systems such as adsorption chromatography, partition chromatography, nonlinear chromatography, and electrophoresis (Kubin, 1965; Kucera, 1965; Vink, 1965; Kaminsikii et al., 1965; Grubner et al., 1966a,b; Grubner et al., 1967; Grubner, 1968; Yamazaki, 1967; Kocisik, 1967; Grushka et al., 1969; Grubner and Underhill, 1970, 1972; Funk and Rony, 1971; Rony and Funk, 1971; Grushka, 1972a; de Clerk and Buys, 1971; Buys and de Clerk, 1972a-f; Villermaux, 1973; Yamaska and Nakagawa, 1973, 1974; Catsimpoolas and Griffith, 1973; Catsimpoolas, et al., 1974).

II. MOMENT ANALYSIS

A. Definition

Before describing the importance of the moments in chromatography their definition should be discussed. The nth statistical moment, m'_n, of a distribution function (or the concentration profile of an eluted peak) $C(t)$ is given by

$$m'_n = \frac{\int t^n C(t)\, dt}{\int C(t)\, dt} \tag{1}$$

The zeroth moment [($n = 0$ in the integral of equation (1)] is simply the normalized area of the peak (in fact, it is unity). The first moment, $n = 1$, is the center of gravity (the mean) of the concentration profile. In chromatography, the first moment is, as will be shown, the retention time of the solute. It will coincide with the peak maximum only if the peak is symmetrical. This is an important point which is neglected by many chromatographers. For precise data handling and analysis, the first moment should be obtained. Moments higher than the first have greater physical significance when they are measured relative to m'_1, i.e.,

$$m_n = \frac{\int (t - m'_1)^n C(t)\, dt}{\int C(t)\, dt} \tag{2}$$

These are called central moments. The second central moment is the peak variance, which is chromatographically related to the plate height H, or the efficiency. The third central moment is indicative of the magnitude and "direction" of the peak asymmetry. Positive values indicate tailing peaks, while negative values point to a fronting shape (typical of sample overloaded chromatogram). The fourth central moment is a measure of the peak flatness as compared to a Gaussian peak. All other higher odd moments have the same physical significance as the third central moment, while the higher even moments provide information similar to the fourth central moment.

In addition, the moments can be used in several expansion series which can approximate the shape of chromatographic peaks. One such series is Gram-Charlier's:

$$C(t) = \frac{1}{\sigma(2\pi)^{1/2}} \exp\left[-\frac{(t - m'_1)^2}{2\sigma^2}\right]\left[1 + \sum_{i=3}^{\infty} \frac{A_i}{i!} H_i\left(\frac{t - m'_1}{\sigma}\right)\right] \tag{3}$$

where t is time, σ^2 is the variance of the peak, $H_i(x)$ is the ith Hermite polynomial, and the coefficients A_i are functions of the moments:

$$A_3 = m_3/m_2^{3/2} \tag{4}$$

$$A_4 = (m_4/m_2^2) - 3 \tag{5}$$

$$A_5 = (m_5/m_2^{5/2}) - 10(m_3/m_2^{3/2}) \tag{6}$$

The coefficients A_3 and A_4 are called skew and excess, respectively. A_3, the skew, measures the asymmetry of the peak, while A_4, the excess, indicates deviation from a Gaussian shape. In the case of Gaussian peaks, all the coefficients A_i in equation (3) vanish and the expansion series is reduced to Gaussian. Mehta *et al.*, (1974), Grubner (1968), as well as McQuarrie (1963) have compared real peaks with the Gram-Charlier approximation. Although the fit is not ideal, the Gram-Charlier series does, indeed, allow one to describe the peak shape. It is, therefore, convenient to discuss not only the moments but also the skew and excess, which are moment-related quantities, owing to the fact that (1) they are coefficients in the expansion series [equation (3)] and (2) they are dimensionless quantities and as such they are independent of the actual physical size of the peak.

The fundamental importance of the moments is thus apparent. Not only do they describe physical quantities, such as retention times, peak width, and asymmetry, but they can be used to approximate the shape of the chromatographic solute zones as they emerge from the system. To complete the picture, the connection between the experimental condition and the moments, skew, and excess is needed. Since the peak shape varies with experimental conditions such as temperature or mobile-phase velocity, the moments must also be a function of these conditions. The desired connection is obtained from a mathematical model describing the chromatographic process, namely, the mass-balance relationship.

B. Mass Balance and the Moments

The mass-balance equations comprise a set of differential equations which attempts to describe the events occurring in the column. The model described here is, mathematically, relatively easy to handle and yet it describes a real chromatographic system. The column, made of a rectangular tubing whose height is much smaller than its width, is coated with a stationary phase of a uniform thickness d_f. The flow profile of the mobile phase is taken to have a plug shape. The assumptions made in this model are described fully elsewhere (Grushka, 1972a.) The system described here is a partition one, as opposed to the adsorption system described by others (Kucera, 1965; Grubner, 1968).

The mass-balance system can be written as follows: In the mobile phase

$$\frac{\partial C_m}{\partial t} = D\frac{\partial^2 C_m}{\partial z^2} - \frac{U\partial C_m}{\partial z} - k_f(KC_m - C_s|x = d_f) \qquad (7)$$

where C_m is the solute concentration in the mobile phase (it is a function of both t and z), t is the time, D is the dispersion coefficient in the mobile phase, z is the longitudinal (column length) axis, U is the carrier velocity, k_f is the mass transfer coefficient across the mobile–stationary phase interface, K is the partition coefficient, and $C_s|x = d_f$ is the concentration of the solute in the stationary phase at the interface of the two phases. The notation $|x = d_f$ denotes values at the mobile–stationary phase interface; x is the lateral coordinator ($x = 0$ at the tube wall). Equation (7) is essentially a modified Fick's second law expression. It states that the rate of change in concentration in an infinitesimally small length unit is a function of (1) solute diffusion in and out of the length element (the first term on the RHS), (2) convective transport by the mobile phase (the second term on the RHS), and (3) the rate of exchange of the solute between the two phases and of the departure from solute-partitioning equilibrium between these phases (the third term on the RHS). The last term on the RHS of equation (7) can be looked upon as the driving force of the chromatographic partitioning and the separation (via K, the partition coefficient). Although thermodynamic forces try to achieve a state of equilibrium between the solute molecules in both phases, the forward movement of the mobile phase ensures that the system is never at the stage of equilibrium.

To relate the surface concentration, $C_s|x = d_f$, to the solute concentration in the stationary phase, the mass-balance equation in that phase is needed.

$$\frac{\partial C_s}{\partial t} = D_s \frac{\partial^2 C}{\partial x^2} \tag{8}$$

where D_s is the solute diffusion coefficient in the stationary phase and x, as mentioned previously, is the lateral coordinate. In writing Equation (8) two assumptions are made: (1) The length element is so small that the surface area of the stationary phase is in contact with a uniform solute concentration in the mobile phase above it. (2) The longitudinal diffusion in the stationary phase element is nil since the concentration gradients in this direction are minute. To complete the system description, initial and boundary conditions are needed:

$$C_m(z, 0) = C_i \quad C_m(\pm\infty, t) = 0 \tag{9}$$

$$C_s(z, 0) = 0 \tag{10}$$

$$\left.\frac{dC_s}{dx}\right|_{x=0} = 0 \tag{11}$$

$$\left.A_s D_s \frac{dC_s}{dx}\right|_{x=df} = V_m k_f (KC_m - C_s)|_{x=df} \tag{12}$$

where A_s and V_m are, respectively, the surface area of the stationary phase and the volume of the mobile phase per unit column volume. A_s can be written as V_s/d_f, where V_s is the volume of the stationary phase per unit column.

The first condition is that at time zero, the injection time, the mobile-phase solute concentration is the initial concentration C_i and at some very long distances from the column inlet the solute concentration is zero. The second condition is that at the injection time, $t = 0$, there is no solute in the stationary phase. Equation (11) is the boundary condition stating that no solute molecules transport through the column wall. Equation (12) is the boundary condition at the phase interface ($x = d_f$) and it describes the exchange of solute molecule across that interface.

In order to solve the above set of differential equations, the mass-balance expression in the stationary phase is solved, and C_s at $x = d_f$ is obtained and put in the mass-balance expression for the mobile phase. Ideally, the complete solution of the concentration profile, $C_m(z, t)$, is desired, but because of mathematical difficulties this is not possible. However, the moments can be obtained from the differential equation, and, in theory, $C(z, t)$ can be approximated via such series as in equation (3). The method by which the differential equations are solved and the moments are obtained is described elsewhere (Grushka, 1972a). Here the first four moments (not including m_0') are given

$$m_1' = \left(\frac{L}{U} + \frac{2D}{U^2}(1 + k) \right) \tag{13}$$

$$m_2 = \left(\frac{2DL}{U^3} + \frac{8D^2}{U^4} \right)(1 + k)^2 + \left(\frac{L}{U} + \frac{2D}{U2} \right)k\left(\frac{2d_f^2}{3D_s} + \frac{V_s}{V_m k_f} \right) \tag{14}$$

$$m_3 = \left(\frac{12D^2L}{U^5} + \frac{64D^3}{U^6} \right)(1 + k)^3 + \left(\frac{DL}{U^3} + \frac{4D^2}{U^4} \right)k(1 + k)\left(\frac{4d_f^2}{D_s} + \frac{V_s}{V_m k_f} \right)$$
$$+ \left(\frac{L}{U} + \frac{2D}{U^2} \right)k\left[\frac{12d_f^4}{15D_s^2} + \frac{4d_f^2 V_s}{D_s V_m k_f} + 6\left(\frac{V_s}{V_m k_f} \right)^2 \right] \tag{15}$$

$$m_4 = \left(\frac{12D^2L^2}{U^6} + \frac{216D^3L}{U^7} + \frac{960D^4}{U^8} \right)(1 + k)^4$$
$$+ \left(\frac{DL^2}{U^4} + \frac{12D^2L}{U^5} + \frac{D^3}{U^6} \right)(1 + k)^2 \left(\frac{8d_f^2}{D_s} + \frac{24V_s}{V_m k_f} \right)$$
$$+ \left(\frac{L^2}{U^2} + \frac{6DL}{U^3} + \frac{12D^2}{U^4} \right)k^2 \left(\frac{4d_f^2}{D_s} + \frac{12V_s}{V_m k_f} \right)$$
$$+ \left(\frac{DL}{U^3} + \frac{4D^2}{U^4} \right)k(1 + k)\left[\frac{32d_f^4}{5D_s^2} + \frac{32d_f^2 V_s}{D_s V_m k_f} + 48\left(\frac{V_s}{V_m k_f} \right)^2 \right]$$
$$+ \left(\frac{L}{U} + \frac{2D}{U^2} \right)k\left[\frac{136d_f^6}{105D_s^3} + \frac{135d_f^4 V_s}{15D_s^2 V_m k_t} \right.$$
$$+ \left. \frac{16d_f^2}{D_s}\left(\frac{3V_s^2}{V_m^2 k_f^2} + \frac{V_s^3}{V_m^3 k_f^3} \right) \right] \tag{16}$$

where k, the capacity ratio, is equal to $K(V_s/V_m)$ and L is the column length. It is to be noted that the moments do, in fact, depend upon such experimental parameters as the column length, mobile-phase velocity, the amount of the stationary phase (via d_f, k, and V_s), etc.

Although the moment expressions are complicated equations, close observation indicates some familiar trends. For example, in the first moment expression, the term $2D/U^2$ is almost always much smaller than L/U. Thus equation (13) can be written as

$$m_1' = (L/U)(1 + k) \tag{17}$$

which is the usual retention-time expression in chromatography. Equation 13, moreover, indicates when equation (17) breaks down. The second central moment, or the peak variance, yields the plate-height expression via

$$H = (Lm_2/m_1')^2 \tag{18}$$

Under normal conditions $2DL/U^3$ is much greater than $8D^2/U^4$ and equation (18) leads to

$$H = \frac{2D}{U} + \frac{2kd_f^2U}{3(1 + k)^2 D_s} + \frac{kV_sU}{(1 + k)^2k_fV_m} \tag{19}$$

This equation is similar to the usual equation of H in open tubular gas chromatography. That the term describing the resistance to mass transfer in the mobile phase is missing from equation (19) is due to the model described here, namely, the approximation of a plug flow of the carrier. The last term on the RHS of equation (19) takes into account the finite rate of solute mass transfer across the mobile–stationary phase interface (James et $al.$, 1964). Under usual operating conditions, it is assumed that $k_f = \infty$. However, in high-speed chromatography this approximation can lead to serious errors. The peak shape can then indicate when the assumption of $k_f = \infty$ is allowed.

The original set of differential equations can be slightly altered to describe a column packed with spherical, solid support particles. If we assume that the dispersion coefficient D in the mobile phase can be written as (Kucera, 1965)

$$D = D_m + AU + BU^2 + \cdots \tag{20}$$

where D_m is the diffusion coefficient of the solute in the mobile phase and A and B are constant, then the plate-height expression for a packed column (assuming $k_f = \infty$) is given by

$$H = A + \frac{2D_m}{U} + BU + \frac{2kd_f^2U}{15(1 + k)^2 D_s} \tag{21}$$

If A is proportional to the diameter of the support particle and B is proportional to the square of the diameter divided by D_m, then the usual van Deemter equation for H results.

The skew and excess can be obtained from the second, third, and fourth central moments. Since the moments do yield expressions which (1) relate to experimental parameters and (2) yield known relationships such as H, it is interesting to see how the skew and excess vary with some experimental conditions. Two examples will be given here: the effect of the column length and effect of k on the skew and excess. These examples are based on the model described earlier and realistic values are given to the parameters which do not change.

1. Effect of Column Length

The following values were arbitrarily given to some of the experimental parameters: $U = 10$ cm/sec, $k = 10$, $D = 0.1$ cm²/sec, $D_s = 1 \times 10^{-5}$ cm²/sec, $d_f = 1 \times 10^{-4}$ cm, and $k_f = \infty$. Figure 1 shows the dependence of skew on the length. It is seen that as L increases, the skew decreases, i.e., the peak becomes more symmetrical as the retention time increases. This is in agreement with the long-time limit of chromatography, which predicts Gaussian peaks for long retention times. At short column lengths the approximation of $k_f = \infty$ is most likely wrong and an actual skew-vs.-length plot might differ somewhat from Figure 1. In any event, Figure 1

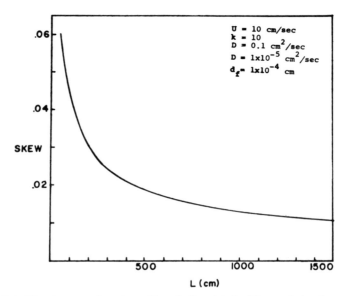

FIGURE 1. The skew as a function of length. Reprinted with permission from *J. Phys. Chem.* **76**: 2586 (1972). Copyright by the American Chemical Society.

has an important message: increasing the column length can improve chromatographic resolution not only because of a higher plate number but also because of increased peak symmetry.

The excess hardly varies with the length. Since the excess is indicative of quantities similar to the plate height, this independence is not surprising. In the example given here, the value of the excess throughout the length range is about 6.32.

2. Effect of the Capacity Ratio

This essentially measures the effect of the partition coefficient on the peak shape. The arbitrary values of the rest of the parameters in equations (14), (15), and (16) are: $U = 10$ cm/sec, $d_f = 1 \times 10^{-4}$ cm, $D = 0.1$ cm^2/sec, $D_s = 1 \times 10^{-5}$ cm^2/sec, $L = 1500$ cm, and $k_f = \infty$. Figure 2 shows the behavior of the excess as a function of k. A maximum in the plot occurs at $k = 1$, in similarity to H vs. k plots for the limiting case where the resistance to mass transfer in the stationary phase is dominant. This well-known observation was studied in detail by dal Nogare and Chiu (1962). The approximation of $k_f = \infty$ is most likely to yield erroneous results at very low k values, where the interface transfer rate might be important. In addition, the plug-flow-profile assumption of the model is most serious at low k owing to the dominance of mass transfer in the mobile phase. Nevertheless, it will be shown shortly that the plot of excess vs. k can be used for solute identification.

Surprisingly, the skew (not shown in Figure 2) does not vary much with k, having a value for the system described here, of about 0.0109 in the k range of 0–50. It might be expected that the skew will decrease with increasing k (long-time limit). However, perhaps because the skew is small here to begin with, its k dependence is insignificant. As will be shown later, the real system shows a larger dependence of k.

The effect of the temperature, the mobile-phase velocity, and the amount of stationary phase are adequately described elsewhere (Grushka, 1972a) and will not be repeated here.

Thus, the moments as well as skew and excess do yield useful information. In fact, the chemical engineers have been using the moments approach to obtain chemisorption rates and the equilibrium constant for adsorption of various gases on different catalysts. In particular, J. M. Smith and his co-workers have utilized the moment approach extensively to obtain transport and kinetic parameters of adsorption by gas chromatography. A recent review by Suzuki and Smith (1975) covers this topic rather thoroughly. As an example, Smith shows that for adsorption chromatography, where the

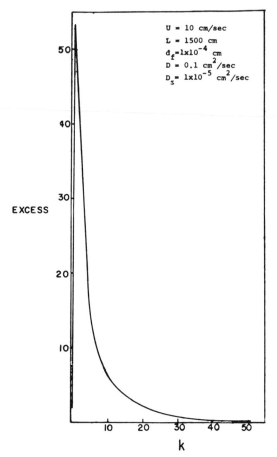

FIGURE 2. The excess as a function of the capacity ratio. Reprinted with permission from *J. Phys. Chem.* **76**: 2586 (1972). Copyright by the American Chemical Society.

column is packed with spherical adsorbents, the first moment can be written as

$$m_1' = \frac{L}{U}[1 + A(1 + BK_a)] \tag{22}$$

where A and B are known functions of the column void volume, particle density, and particle porosity, and K_a is the adsorption equilibrium constant. Thus, from a plot of m_1' vs. L/U adsorption equilibrium constants can be obtained. Indeed, Smith found that for silica gel adsorbent and several hydrocarbon solutes the agreement between K_a values obtained from gas chromatography and from the BET method is excellent.

C. Other Uses of Moments Analysis

In addition to the above discussion, the moment can also be used for the recognition of strongly overlapped peaks (Grushka *et al.*, 1970). As an example, assume initially that the peaks are Gaussian. For a single peak the skew and the excess are zero. However, in the case of strongly overlapped peaks, i.e., resolution R_s less than 0.5, the excess will not equal zero. [The resolution is defined as $2\Delta t_R/(W_1 + W_2)$, or the difference in retention time divided by half the peak-width sum.] This is shown in Figure 3, which is a composite of two Gaussians of equal height and width. Under normal conditions, closely eluting peaks have almost equal width. However, the equal-height system shown in Figure 3 was chosen for illustrative purposes only. Because the two peaks are equal, the skew is zero at any resolution level. The excess, on the other hand, decreases as the resolution increases. Thus, at resolutions of 0.25, even though the composite looks like a single peak, the negative excess indicates existence of double peaks. If the heights of the two peaks in the composite were not equal, the skew would also be nonzero (positive if the more retained peak is the smaller and negative if it is the larger). In a Gaussian world, the recognition of strongly overlapped peaks would be a simple task: nonzero skew and excess would be indicative of double peaks.

In practice, however, chromatographic peaks are not Gaussian and a more realistic peak shape is described by an exponentially modified Gaussian curve. Such a peak looks like Gaussian at its beginning, but, once past its

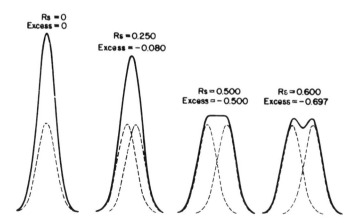

FIGURE 3. Excess of two equal and overlapping Gaussian peaks. Reprinted with permission from *Anal. Chem.* **42**: 21 (1970). Copyright by the American Chemical Society.

maximum, it begins to tail. The magnitude of the tailing depends upon the experimental conditions. Most of the tailing occurs as a result of extracolumn effects such as dead volume in the connecting tubing, detector, and injector and the time constant associated with the electronics. An excellent review of extracolumn effects is given by Steinberg (1966). More recently, McWilliam and Bolton (1969, 1971) have also demonstrated how extracolumn effects lead to an exponentially modified Gaussian peak.

An exponentially modified Gaussian curve, i.e., a Gaussian curve convoluted with an exponential decay function, is given by

$$C(t) = \frac{A}{2\sigma(2\pi)^{1/2}} \int_0^+ \exp\left[\frac{-(t - t_R - t')^2}{2\sigma^2}\right] \exp\left(\frac{-t'}{\tau}\right) dt' \qquad (23)$$

where A is the peak amplitude, t_R is the center of gravity of the Gaussian peak, σ^2 is the variance of the Gaussian peak, τ is the time constant of the exponential modifier, and t' is a dummy variable of integration. Also, τ is a function of the various extracolumn contributor. The asymmetry of the peak depends upon the ratio τ/σ. Figure 4 shows three traces: for $\tau/\sigma = 0$, $\tau/\sigma = \infty$, and $\tau/\sigma = 1.5$, i.e., pure Gaussian, pure exponential decay, and moderately tailing peaks, respectively.

The moments of a single exponentially modified peak were obtained by Grushka (1972b):

$$m_1' = t_R + \tau \qquad (24)$$

$$m_2 = \sigma^2 + \tau^2 \qquad (25)$$

$$m_3 = 2\tau^3 \qquad (26)$$

$$m_4 = 3\sigma^2 + 6\sigma^2\tau^2 + 9\tau^4 \qquad (27)$$

$$\text{Skew} = \frac{2(\tau/\sigma)^3}{[1 + (\tau/\sigma)^2]^{3/2}} \qquad (28)$$

$$\text{Excess} = \frac{3[1 + (2\tau^2/\sigma^2) + (3\tau^4/\sigma^4)]}{[1 + (\tau/\sigma)^2]^2} - 3 \qquad (29)$$

It is interesting to note that the first three moments are simply the superposition of the moments of the Gaussian and of an exponential decay. The third moment in theory allows the measurements of the time constant of a system, provided the column itself produces Gaussian peaks.

The behavior of the skew and excess as a function of τ/σ is shown in Figure 5. The plot extends only to the value of $\tau/\sigma = 5$ since past that value the change in the skew and excess are small. Figure 5 is important since it can indicate the value of τ which begins to severely affect the peak shape. For example, if a skew of only 0.1 can be tolerated, Figure 5 shows that τ/σ

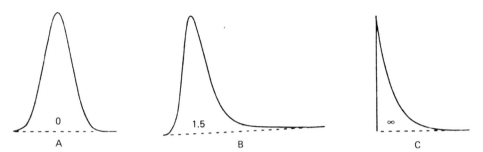

FIGURE 4. Some exponentially modified Gaussian peaks: (A) $\tau/\sigma = 0$, (B) $\tau/\sigma = 1.5$, (C) $\tau/\sigma = \infty$.

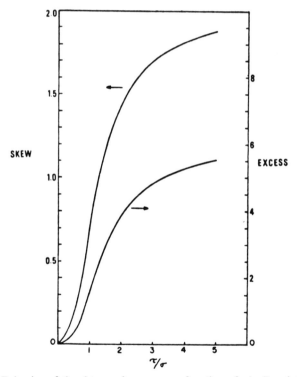

FIGURE 5. Behavior of the skew and excess as a function of τ/σ. Reprinted with permission from *Anal. Chem.* **44**: 1733 (1972). Copyright by the American Chemical Society.

cannot exceed 0.4. Thus, if the total time constant of the system exponential decay is 0.1 sec, the chromatographic zone broadened by the column must have $\sigma \geq 0.25$ sec (or a base width of 1.0 sec since 4σ equals peak width). Narrower peaks will be distorted to a much greater extent (skew > 0.1). If $\tau = 1$ sec, the peak width can be as narrow as 10 sec before serious distortion occurs. Thus, the moments, the skew, and the excess, in conjunction with plots such as those in Figure 5, can indicate the permissible limits of extra-column contributions.

The skew and excess, if they are to be viable peak indicators, must be able to deal with peak shapes such as the one just described. In Figure 6, the excess is plotted vs. skew. The solid line corresponds to the skew and excess of a single peak at various τ/σ values. As τ/σ increases from zero, so do the skew and excess. The moments, skew, and excess of overlapped exponentially modified Gaussian shapes were found (Grushka, 1972b), and their theoretical values as a function of the peak separation d were computed.

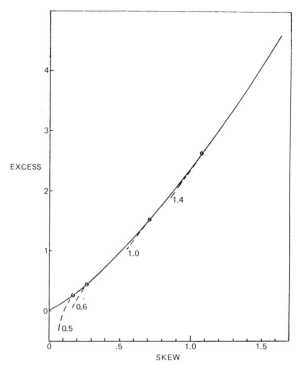

FIGURE 6. Excess vs. skew. Each dashed line corresponds to a system of two identical peaks. The number next to each dashed line is the τ/σ value of the two peaks in the composite. Reprinted with permission from *Anal. Chem.* **44**: 1733 (1972). Copyright by the American Chemical Society.

Each dashed line in Figure 6 belongs to a composite of two equal (in height and width) peaks at a particular value of τ/σ. That value is indicated next to each dashed line. Each line extends to a separation of about 1.4 σ units. It is seen that (1) the closer the peaks are to Gaussian (low τ/σ values) the easier is the double-peak recognition and (2) the larger the separation, the easier is the recognition. At zero separation, since the two peaks in the composite are equal, the skew and excess are equal to that of a true single peak. Hence all the dashed lines start at the single-peak line (the solid line in Figure 6). Note that all the dashed lines, i.e., the skew–excess curve for double peaks, fall below the single-peak solid lines. However, at τ/σ values of about 1 and above the double-peak lines are close to the solid line, and the recognition might be difficult for two identical peaks even at separations of 1.4 σ units ($R_s \simeq 0.35$).

Two closely eluted peaks most likely will have almost identical τ/σ values. Their heights, on the other hand, frequently are not equivalent. To investigate the effect of the height, the skew and excess of a composite made of two peaks each with τ/σ of 1.4 were studied. The value of 1.4 for τ/σ gives realistic peak shapes (very slight tails). Figure 7 shows an excess-vs.-skew plot for such double-peak systems. Each line corresponds to a different peak-height ratio as indicated in the figure. The peak heights of the first and second (more retained) peaks are indicated by A and B, respectively. The height ratio of the first to the second peak varies from 0.2 to 5. All the lines begin at the same point, i.e., that of a single peak with $\tau/\sigma = 1.4$, and each line extends to a separation d of 1.4 σ units ($R_s \simeq 0.35$). It is seen that each line is sufficiently different from the equal-height line. The change in the skew and excess is most pronounced as the height ratio increases. This is to be expected since as one of the peaks becomes smaller, the overall perturbation in the shape decreases, and in the limit of a vanishingly small peak the doublet becomes a singlet. Note that the lines in Figure 7 are not symmetrical around the equal-height line. This is due to the fact that each of the peaks is asymmetrical and the position of the peaks does affect the shape of the composite. Since the lines in Figure 7 do not cross one another, it can be inferred that in this coordinate system for two peaks of $\tau/\sigma = 1.4$ there is a uniquely defined region for double peaks of various height ratios. Each point in this region describes two overlapped peaks of a certain resolution and a certain height ratio. As expected, as the resolution decreases, the double-peak recognition becomes more difficult.

A comparison of Figures 6 and 7 shows that some of the lines corresponding to the leading peak, being smaller, lie above the solid line. In general, it can be said that points below or above the solid line belong definitely to double peaks; those points above consist of a smaller leading peak while the points below describe systems where the first peak can be either the taller or the shorter. Unfortunately, the solid line in Figure 6 is not uniquely defined

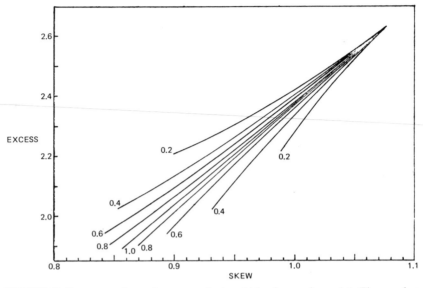

FIGURE 7. Excess vs. skew of a composite in which $\tau/\sigma_1 = \tau/\sigma_2 = 1.4$. The number next to each line indicates the height ratio of the two peaks. The numbers above the 1.0 line (corresponding to $A = B$ are A/B values; numbers below that line are B/A values. Reprinted with permission from *Anal. Chem.* **44**: 1733 (1972). Copyright by the American Chemical Society.

and points on it can belong to either single or double peaks. This is particularly true for systems with large τ/σ values.

It should be noted that in theory, if τ is known, then plots such as those in Figure 7 allow complete characterization of the system, i.e., the height ratio as well as the separation of the peaks can be found. The same analysis can be made in the case where the second peak is slightly wider than the first one.

Studies of other peak-shape models such as bi-Gaussian, Poisson, and kinetic tailing (Grushka *et al.*, 1970) also indicated the feasibility of using skew and excess for the discernment of double peaks.

III. EXPERIMENTAL STUDIES

A. Verification of Skew and Excess Utility

Experimentally, the moments are obtained by digitizing the chromatographic detector signal and manipulating the data with the aid of a computer (or even a programable calculator). The moments are calculated from the expression

$$m_1' = 1(/A) \sum y_i(t_i - 0.5\Delta t) \tag{30}$$

where y_i is the peak height at the time t_i that the digitization occurred, Δt is the interval between two successive digitizations, and A is the peak area. The central moments then are

$$m_n' = (1/A) \sum y_i(t_i - 0.5\Delta t - m_1')^n \tag{31}$$

The factor $0.5\Delta t$ takes into account the finite acquisition time between the data points.

The experimental difficulties in obtaining the third and higher central moments should not be underestimated. Chesler and Cram (1971, 1972) have discussed in detail the various parameters which affect the accuracy of the experimentally obtained statistical moment.

In addition to the experimental system described by us (Grushka *et al.*, 1969, 1971), Overholtzer and Rogers (1969) described a high-precision setup for moment measurements.

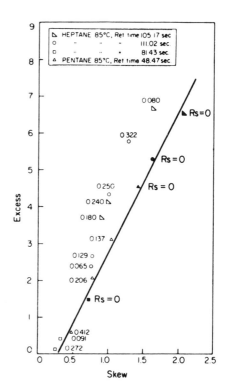

FIGURE 8. Excess vs. skew for pentane and heptane. The resolution is shown for each point. Reprinted with permission from *Anal. Chem.* **42**: 21 (1970). Copyright by the American Chemical Society.

To simulate double peaks, an injection valve was used to inject the solute rapidly twice. The peak separation was determined from the time interval between the two successive injections. The resolution was then determined from the usual chromatographic relationship.

Figure 8 shows the results for some alkenes at various temperatures. The single peaks are shown in bold type. The resolution of each doublet is indicated next to each point. It is seen that the skew and excess of the double peaks are sufficiently different from those of single peaks. In fact, most of the overlapped-peaks data fall to the left of a line which connects the single-peak points. In general, as the resolution increases, the skew and excess decrease.

Figure 9 shows a double peak simulated with methane. Again, it is seen that the skew and excess of double peaks are different than those of the single peak, and the same trend as in Figure 8 occurs. Similar results were obtained with ketones and alkenes (Grushka *et al.*, 1970).

It is thus seen that the skew and excess can differentiate between double and single peaks. The technique can be particularly powerful if a standard

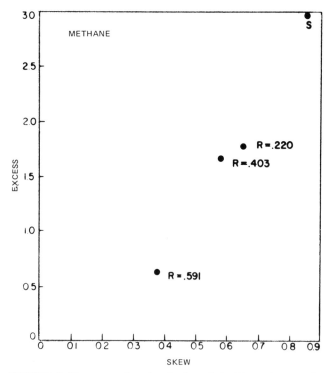

FIGURE 9. Excess vs. skew for single and double methane peaks.

chromatogram is available, and in this case, a comparison between the skew and excess of the standard and the unknown should reveal the existence of multiple peaks.

B. Peak Identification

In the previous studies, it was noted that the skew and excess of several homologous series fall in particular regions when plotted in a skew-excess coordinate system. This effect might be expected since, as mentioned, the peak shape, of which the skew and excess are a measure, depends on the processes occurring in the chromatographic column. The shape can then be used in peak identification. The idea of using peak shape as a qualitative indicator is certainly not new since, as most practicing chromatographers know, peak tailing has been used, in a purely empirical manner, as an indication of polar groups. Riedmann (1969) suggested that often the peak shape might identify different classes of compounds. No systematic approach, however, was carried out to study the relationship between the solute peak shape and nature. We will next describe some initial results indicating the feasibility of using the skew and excess as peak identifiers.

There exist many identification schemes, some of which are specific and some of which are general to a particular class of compounds (Leathard and Sherlock, 1970). In general, these techniques are based on retention data or on post- or precolumn instrumentation. In the former case, ambiguity in the retention data may necessitate more than one chromatographic run. In the latter, the ancillary equipment tends to be expensive. The peak shape, on the other hand since it contains all the chromatographic information that can be obtained, should be able to provide directly information pertaining to the nature of the solute. The following will demonstrate that this approach can, indeed, be beneficial.

Since the skew and excess are velocity and temperature sensitive, all our work was carried at 51.3°C and at a constant (to within 0.25%) average velocity of 19.4 cm/sec. The digitizing rate was so adjusted as to fit the criteria set by Chesler and Cram (1971). McNair and Cook (1972) indicated a strong sample-size dependence of the skew and excess, therefore, we were careful not to overload the column capacity.

Grushka *et al.* (1969) demonstrated the importance of the position of the digitizer baseline with respect to that of the electrometer. Therefore, the baseline of the electrometer was continuously monitored with a digital voltmeter (DVM) and compared with the output of the digitizier. The digitizing process was stopped when the DVM indicated that the peak was completely eluted and that the previously set electrometer baseline was re-established. With these procedures, the precision in obtaining the skew of each solute was at worst 4%, and typically 1–2%. The excess of each solute was obtained with a precision of at worst 8% and typically 4%.

The data of five homologous series are shown in Figure 10. Except with the alkanes, three members of each series were investigated. With most homologous series, solutes with higher boiling points had in them impurities which were eluted too close to the main peak. More on the problem of impurities will be discussed later.

Figure 10 indicates clearly that, at least for the systems which we investigated, different homologous series tend to "congregate" in different regions of the coordinate system. On Apiezon L (the stationary phase used), the polarity of the alcohols is well demonstrated by their high skew and excess values. On the other hand, alkanes and symmetrical aromatics are characterized by lower skew and excess. The behavior within each series, unfortunately, and unexplainable at this point, is not uniform, e.g., the skew and the excess do not necessarily increase with the boiling points or with the carbon number.

Figure 10 shows that some solutes belonging to different homologous series have either similar skew or similar excess. This phenomenon can, conceivably, cause misinterpretation of the data. However, as Figure 10 shows, if the skews of the two solutes are identical, or close in magnitude, the excess values are different, thus removing possible ambiguity; the inverse is also true.

In addition to the skew and excess, one has another parameter at his disposal; namely, the retention time or retention volume (more appropriately it should be the first moment). This is related to the dependence of the skew and excess on the capacity ratio as previously described. The retention time

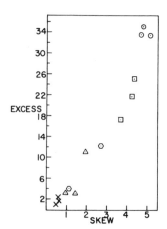

FIGURE 10. Skew vs. excess of several homologous series: (×) aromatics, (△) alkanes, (○) alkenes, (□) primary alcohols, (○) secondary alcohols.

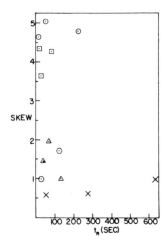

FIGURE 11. Skew vs. retention time (t_R): (×)
aromatics, (△) alkanes (○) alkenes, (□) primary
alcohols, (○) secondary alcohols.

is a function of the capacity ratio. Figures 11 and 12 are, respectively, plots
of the skew vs. the retention time and the excess vs. the retention time. With
use of these three parameters, all ambiguities can be removed. Figure 10
shows a case in point, the aromatic solutes. All the skew and excess values
for the three solutes are bunched together. Figures 11 and 12, however,
clearly allow the differentiation of the different species. Similarly, take the
case of benzene and 3-n-pentenol. The retention time of these two solutes
are very close to one another (53.1 sec and 52.7 sec, respectively). If one uses
retention times as means of identification, an error could result. Figures 11
and 12 (and Figure 10 for that matter), clearly removes the possibility of
erroneous misinterpretation.

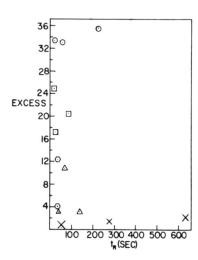

FIGURE 12. Excess vs. retention time
(t_R): (×) aromatics, (△) alkanes, (○)
alkenes, (□) primary alcohols, (○) secondary alcohols.

The latter point has a very important practical ramification. To date, the problem of similar retention times for different solutes was dealt with by either utilizing two columns, each with a stationary phase of different polarity, or, if the thermodynamics properties of the solutes are different, by running the analysis at two different temperatures (Leathard and Sherlock, 1970). The skew and excess in conjunction with the retention time allow, in essence, solute identification on one column at a single temperature. The possibility that three of these parameters will coincide is rather remote since, as mentioned previously, the shape is a reflection of the processes occurring in the column. Hence, chemically similar solutes might have close values for their skew and excess but their retention times will differ. Conversely, chemically unrelated substances might have similar retention times but owing to the differences in their natures they will, most likely, have different skew and excess values.

A drawback which exists with most other identification schemes is the necessity of preparing calibration curves. An additional—and perhaps unique to this method—disadvantage is the adverse effect of small amounts of impurities which are not well separated from the peak of interest. The skew and, particularly, the excess are rather sensitive to contamination [or to noise for that matter (Chesler and Cram, 1971)]. However, in many instances, the peaks of interest are well separated from any impurity, in which case the skew–excess method should work without difficulties.

The problem of a small impurity on a major peak can, in some cases, be a blessing in disguise. For example, in process and quality control one might be interested in monitoring just that situation. Here, however, the important criterion might be the amount rather than the nature of the impurity. Again, with the aid of calibration curves, the amount of the minor component can be ascertained from the changes in skew and excess of this composite peak.

The skew and excess analysis will not replace some of the existing identification techniques. Obviously, the gas chromatography/mass spectrometry system, for example, is a powerful analytical method. Also, the retention data can be adequately used in many instances. Moment analysis is meant to be a complementary rather than a replacement method for identifying chromatographic elements.

IV. SLOPE ANALYSIS

In addition to moment analysis, peak shape can be studied by investigating its slope. To date, slope analysis utilizes mainly first derivatives (Giese and French, 1955; McWilliam, 1969; Ashley and Reilly, 1965;

Kambara and Saitah, 1968). The second derivative has been used mainly for location of the peak's maximum in a moderately overlapped system. Here it will be shown how strongly overlapped peaks can be recognized by proper manipulation of second-derivative information. Slope analysis is simpler than moment analysis, both conceptually and mathematically.

As with moment analysis, a study of a Gaussian peak is a suitable starting point. The second derivative of a single Gaussian peak has two maxima and one minimum. The magnitudes of these extrema depend upon the variance and the height of the peak. However, the ratio between the extrema, for a single Gaussian curve, is independent of these parameters. In fact, the ratio of either maximum to the minimum is $-2 \exp[-3/2]$, and the ratio between the two maxima is unity. It is best here to define some ratio parameters which will be used shortly. The ratio of the maximum in the second-derivative curve on the leading side of actual peak maximum to the minimum will be indicated by R_1; the ratio of the maximum on the back side of the peak to the minimum will be indicated by R_2; and the ratio of the two second-derivative maxima will be R_3.

In the case of a strongly overlapped Gaussian peak where the composite appears as a single envelope, the second derivative will still have two maxima and one minimum, e.g., Figure 13. The method of obtaining the plot is given elsewhere (Grushka, and Monacelli, 1972) As Figure 13 shows, the ratios of

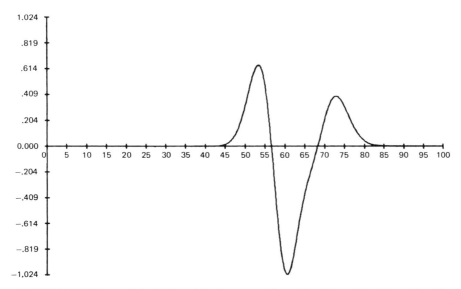

FIGURE 13. A second-derivative plot of a composite made of two Gaussian peaks of equal width but unequal heights.

the two maxima to the minimum are no longer equal, and therein lies the found-
ation for double-peak recognition. It can be asked why not use the first derivative, where the ratio of the maximum to the minimum will also differ from
unity in the case of unequal overlapped peaks. The reasons are that (1) the
first-derivative data are not sensitive enough when one of the peaks in the
composite is 0.1 the height of the second, (2) the first-derivative ratio is
always unity in the case of two equal peaks, and (3) the first-derivative data
have only one diagnostic parameter, namely, the maximum-to-minimum
ratio. The second-derivative data do not suffer from these limitations. The
existence of double peaks can be recognized even if the two peaks are identical
(Grushka, and Monacelli, 1972). Higher derivatives can yield more information; however, experimental difficulties due to noise magnification limit one to
the second derivative. It should be noted here that when the second-derivative
curve has more than three extrema, the usefulness of the method is fortuitous
since this indicates a break in the actual chromatographic peak (either as
a shoulder or as a shallow valley) and visual inspection will indicate the
existence of double peaks.

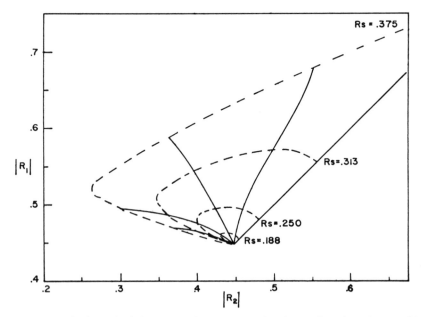

FIGURE 14. $|R_1|$ vs. $|R_2|$ for composite peaks made of two Gaussian shapes with
$\sigma_1 = \sigma_2$. The peak-height ratios, B/A, for each solid line, from right to left, are 1, 0.75,
0.50, 0.20, and 0.10. Dashed lines indicate contours of constant R_S. Reprinted with permission from *Anal. Chem.* **44**: 484 (1972). Copyright by the American Chemical Society.

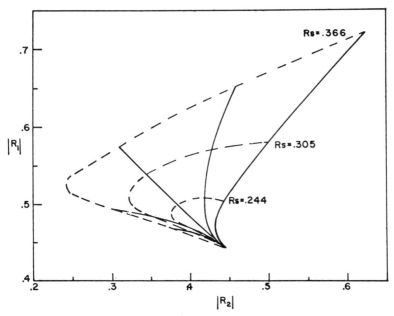

FIGURE 15. $|R_1|$ vs. $|R_2|$ for composite peaks made of two Gaussian shapes with $\sigma_2 = 1.05\sigma_1$. The peak-height ratios, B/A, for each solid line, from right to left, are 1, 0.75, 0.50, 0.20, and 0.10. Dashed lines indicate contours of constant R_s. Reprinted with permission from *Anal. Chem.* **44**: 484 (1972). Copyright by the American Chemical Society.

To check the usefulness of slope analysis, double peaks were simulated at different resolution levels (up to $R_s \simeq 0.375$) and at different height ratios. Figure 14 shows a family of curves, each belonging to a composite made of two peaks with $\sigma_1 = \sigma_2$ (i.e., equal width) and of various first-to-second peak-height ratios (indicated in the figure caption). Each line extends up to a resolution of about 0.375. The coordinate system is the absolute value of R_1 vs. the absolute value of R_2. It is immediately clear from Figure 14 that in this coordinate system (which is similar to the skew–excess system) there is a region which is uniquely defined for a particular composite peak of a certain height ratio at a certain resolution. Each point in this domain belongs to a particular set of double peaks with a unique height ratio and unique resolution.

Note that all the lines emerge from a single point, the $|R_1|$ and $|R_2|$ values of a single Gaussian peak (irrespective of its height or width). This is so because when two Gaussian peaks merge completely (resolution of 0) the composite is also Gaussian. As expected, the smaller the second peak the harder is the double-peak recognition. Also, at low resolutions it is difficult to discern the existence of multiple peaks. The broken lines are contours of constant resolution since they connect all the points of equal resolution for any peak-height ratio.

Figure 14 is one in which the first peak is largest. When the second peak is taller, the lines in Figure 14 will be reflected about the equal-peaks line (where $B/A = 1$).

Figure 15 shows the situation where the second peak is somewhat wider than the first one, i.e., $\sigma_2 = 1.05\sigma_1$. Again, it is seen that there is a region which uniquely defines the double peaks, their resolution, and their height ratio.

As mentioned before, real chromatographic peaks may be better approximated by the exponentially modified Gaussian form given in Equation (23). The method of finding the slope of this convolution integral is given elsewhere (Grushka, 1972b). The ratios R_1 to R_2 will, of course, depend on the τ/σ value, and in Figure 16, the solid line is generated when R_1 is plotted vs. R_2 for various τ/σ values. The line starts at the point where $R_1 = R_2 = 0.446$, which corresponds to a true Gaussian curve. As τ/σ increases, R_1 increases while R_2 decreases. This is to be expected since as the τ/σ ratio increases

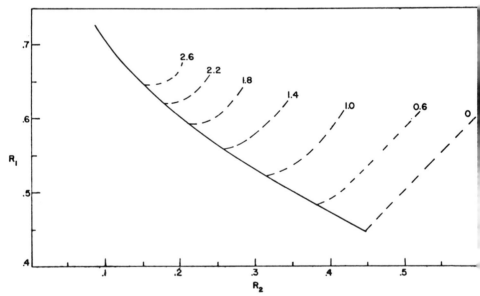

FIGURE 16. R_1 vs. R_2. Each dashed line corresponds to a system of two identical peaks. The number next to each dashed line is the τ/σ value of the two peaks in the system. Reprinted with permission from *Anal. Chem.* **44**: 1733 (1972). Copyright by the American Chemical Society.

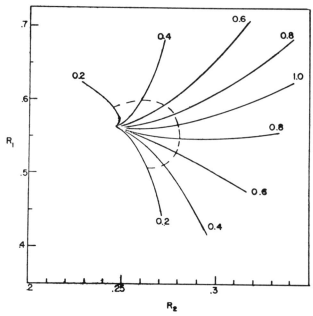

FIGURE 17. R_1 vs. R_2 of a composite in which $\tau/\sigma_1 = \tau/\sigma_2 = 1.4$. The number next to each line indicates the height ratio of the two peaks. The numbers above the 1.0 line (corresponding to $A = B$) are B/A values; numbers below that line are A/B values. Reprinted with permission from *Anal. Chem.* **44**: 1733 (1974). Copyright by the American Chemical Society.

the front end of the peak sharpens while the back end flattens. This single-peak line is extended only to the point where τ/σ is about 5.

The dashed line in Figure 16 shows the behavior of R_1 and R_2 for several composites, each made of identical peaks of a given value of τ/σ value. Each dashed line begins at a separation, d, of zero, where the two peaks completely overlap, and extends to a separation of 1.4 σ units. At $d = 0$, since the two peaks are identical, the composite is equivalent to a single peak. Thus, all the dashed lines begin at the solid line. If the two peaks in the composite are identical, then, as Figure 16 shows, the recognition of double peaks is easy. In practice, of course, the height ratio of the peaks in the composite is not necessarily unity. To investigate the effect of the relative peak height attention was centered on the same system discussed in Section II, namely $\tau/\sigma_1 = \tau/\sigma_2 = 1.4$.

Figure 17 shows the behavior of R_1 and R_2 of the above-mentioned composite system for several peak-height ratios. Each solid line extends from $d = 0$ to $d = 1.4$ σ units. All the lines emanate from a single point, namely, a single peak with $\tau/\sigma = 1.4$. It is evident that a region which is uniquely defined exists in this coordinate system. Each point in that region defines the height ratio of the two peaks as well as their separation. To further demonstrate the usefulness of the plot, the dashed line in Figure 5 is a contour of constant d; in this case $d = 1\ \sigma$ unit. Any point on it belongs to a composite peak made of two exponentially modified Gaussian shapes with $\tau/\sigma = 1.4$ at a unique height ratio. As expected, when one of the peaks in the composite is very short or at very low resolution levels, the characterization is more difficult.

As in the case of moment analysis, due to the asymmetrical nature of the model, the values of R_1 and R_2 are not symmetrical around the equal-heights line. Also, a composite made of peaks of unequal height (regardless which is the largest) might cross the solid line; thus, points on that line unfortunately can belong to either a single or double peak.

The same behavior occurs when the second peak is slightly wider (Grushka, 1972b).

Experimental results are shown in Figures 18 and 19. Again, an injection valve was used for simulation of double peaks. In addition, the valve allowed

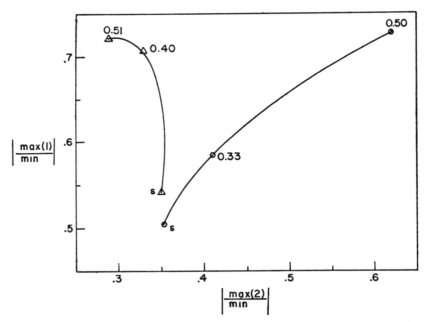

FIGURE 18. $|R_1|$ vs. $|R_2|$ for composites made of two methane peaks. Resolution between the peaks is given next to each experimental point. Here s signifies single peaks. ($\odot$) Peak heights in the composite identical, ($\triangle$) $B/A = 0.5$. Reprinted with permission from *Anal. Chem.* **44**: 484 (1972). Copyright by the American Chemical Society.

injection of different amounts of the solute so that unequal peaks could be
studied. Experimentally, it was easier to simulate the case where the first
peak was the larger one. The resolution between the two peaks in the composite was controlled by the time interval between consecutive injections.
The digitized data were manipulated by the computer to give a smoothed
second derivative directly from the same data (Grushka and Monacelli,
1972).

$$Y'' = \frac{1}{7\Delta t^2}(2Y_{i-2} - Y_{i-1} - 2Y_i - Y_{i+1} + 2Y_{i+2}) \qquad (32)$$

where Y_i is the peak height at the ith interval and Δt is the time interval of
digitization.

The experimental data show that slope analysis can be used for double-
peak recognition. The R_1 and R_2 values of double peaks are different from
those of single peaks. In addition, as in the theoretical section where the
second peak (the more retained one) is smaller, the curve shifts to the left

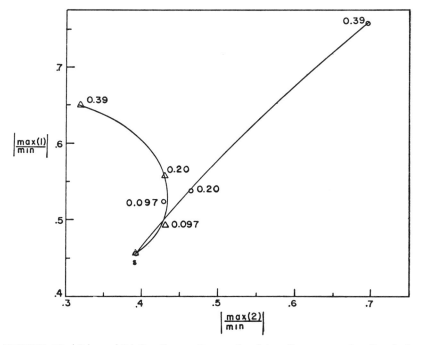

FIGURE 19. $|R_1|$ vs. $|R_2|$ for Composites made of two heptane peaks. Resolution
between the peaks is given next to each experimental point. Here s signifies a single
peak. ($\odot$) Peak heights in the composite identical, ($\triangle$) $B/A = 0.5$. Reprinted with
permission from *Anal. Chem.* **44**: 484 (1972). Copyright by the American Chemical
Society.

of the equal-peaks line. It should be mentioned that double peaks made of double injections of p-xylene were also studied (Grushka and Monacelli, 1972). In this case, the equal-peaks line was anomalous in that R_2 increased while R_1 decreased as the resolution increased. But even in this case, the recognition of double peaks was easily accomplished.

This method of analysis can be useful in routine analysis such as quality and process control. The data treatment is simple and can be handled by a microcomputer. The fact that the exact peak shape may be unknown does not limit the effectiveness of the method provided that calibration curves are available. The calibration of the system can be done in two forms. If only the recognition of double peaks is desired, then the calibration is done simply by obtaining R_1 and R_2 of one of the components suspected to be in the composite. For quantitation purposes, the calibration should be made with various amounts of the two solutes known to be in the composite. This latter method should prove most useful in the case where the constituents in the mixture are known, i.e., process control.

V. CONCLUSION

It is seen that moment and slope analysis can be useful in peak characterization. Moments analysis is more powerful simply because the moments relate to the physical (and chemical) parameters affecting the peak shape. From the moments, diverse physicochemical data can be obtained. In addition, the moments can indicate the courses of action that should be taken to improve the chromatographic separation.

Both moment and slope analysis can be utilized for double-peak recognition. There, slope analysis might be the method of choice since it is conceptually and mathematically the easier.

VI. REFERENCES

Ashley, J. W. and Reilly, C. N. (1965) *Anal. Chem.* **37**: 626.
Bocke, J. and Parke, N. G., III (1962) in *Gas Chromatography* (N. Brenner, ed.) p. 391, Academic Press, New York.
Buys, T. S. and de Clerk, K. (1972a) *J. Chromatogr.* **67**: 1.
Buys, T. S. and de Clerk, K. (1972b) *J. Chromatogr.* **67**: 13.
Buys, T. S. and de Clerk, K. (1972c) *Sep. Sci.* **7**: 441.
Buys, T. S. and de Clerk, K. (1972d) *Sep. Sci.* **7**: 543.
Buys, T. S. and de Clerk, K. (1972e) *Anal. Chem.* **44**: 1273.
Buys, T. S. and de Clerk, K. (1972f) *J. Chromatogr. Sci.* **10**: 722.

Catsimpoolas, N. and Griffith, A. L. (1973) *Anal. Biochem.* **56**: 100.
Catsimpoolas, N., Yotis, W. W., Griffith, A. L., and Rodbard, D. (1974) *Arch. Biochem. Biophys.* **163**: 113.
Chesler, S. N. and Cram, S. P. (1971) *Anal. Chem.* **43**: 1922.
Chesler, S. N. and Cram, S. P. (1972) *Anal. Chem.* **44**: 2240.
dal Nogare, S. and Chiu, J. (1962) *Anal. Chem.* **34**: 890.
de Clerk, K. and Buys, T. S. (1971) *J. Chromatogr.* **63**: 193.
Funk, J. E. and Rony, P. R. (1971) *Sep. Sci.* **6**: 365.
Giddings, J. C. (1965) *Dynamics of Chromatography*, Marcel Dekker, New York, New York.
Giese, A. T. and French, C. S. (1955) *Appl. Spectrosc.* **9**: 78.
Glueckauf, E. (1955) *Trans. Faraday Soc.* **51**: 34.
Grubner, O. (1968) in *Advances in Chromatography* (R. A. Keller and J. C. Giddings, eds.) p. 173, Marcel Dekker, New York.
Grubner, O. (1971) *Anal. Chem.* **43**: 1934.
Grubner, O. and Underhill, D. W. (1970) *Sep. Sci.* **5**: 555.
Grubner, O. and Underhill, D. W. (1972) *J. Chromatogr.* **73**: 1.
Grubner, O., Ralek, M., and Zikenova, A. (1966a) *Collect. Czech. Cohem. Commun.* **31**: 852.
Grubner, O., Ralek, M., and Kucera, E. (1966b) *Collect. Czech. Chem. Commun.* **31**: 2629.
Grubner, O., Zikenova, A., and Ralek, M. (1967) *J. Chromatogr.* **28**: 209.
Grushka, E. (1972a) *J. Phys. Chem.* **76**: 2586.
Grushka, E. (1972b) *Anal. Chem.* **44**: 1733.
Grushka, E. and Monacelli, G. C. (1972) *Anal. Chem.* **44**: 484.
Grushka, E., Myers, M. N., Schettler, P. D., and Giddings, J. C. (1969) *Anal. Chem.* **41**: 889.
Grushka, E., Marcus, M. N., and Giddings, J. C. (1970) *Anal. Chem.* **42**: 21.
James, A. T. and Martin, A. J. P. (1952) *Biochem. J.* **50**: 679.
James, M. R., Giddings, J. C., and Eyring, H. (1964) *J. Phys. Chem.* **68**: 725.
Kambara, T. and Saitah, K. (1968) *J. Chromatogr.* **35**: 318.
Kaminiskii, V. A., Timoshev, S. F., and Tunitskii, N. N. (1965) *Russ. J. Phys. Chem* **39**: 1354.
Kocisik, M. (1967) *J. Chromatogr.* **30**: 459.
Kubin, M. (1965) *Collec. Czech. Chem. Comun.* **30**: 1104.
Kucera, E. (1965) *J. Chromatogr.* **19**: 237.
Lapidus, L. and Amundson, N. R. (1952) *J. Phys. Chem.* **56**: 984.
Leathard, D. A. and Sherlock, B. C. (1970) in *Identification Techniques in Gas Chromatography*, Wiley-Interscience, New York.
McNair, H. M. and Cooke, W. M. (1972) *J. Chromatogr. Sci.* **10**: 27.
McQuarrie, D. A. (1963) *J. Chem. Phys.* **38**; 437.
McWilliam, I. G. (1969) *Anal. Chem.* **41**: 674.
McWilliam, I. G. and Bolton, H. C. (1969) *Anal. Chem.* **41**: 1755.
McWilliam, I. G. and Bolton, H. C. (1971) *Anal. Chem.* **43**: 883.
Martin, A. J. P. and Synge, R. L. M. (1941) *Biochem. J.* **35**: 1358.
Mehta, R. V., Merson, R. L., and McCoy, B. J. (1974) *J. Chromatogr.* **88**: 1.
Nielson, K. L. (1964) in *Methods in Numerical Analysis*, 2nd ed., Macmillan, New York.
Oberholtzer, J. E. and Rogers, L. B. (1969) *Anal. Chem.* **44**: 1234.
Riedmann, M. (1969) *Ber. Bunsenges. Phys. Chem.* **69**: 840.

Rony, P. R. and Funk, J. E. (1971) *Sep. Sci.* **6**: 383.

Steinberg, J. C. (1966) in *Advances in Chromatography* (J. C. Giddings and R. A. Keller, eds.) p. 205, Marcel Dekker, New York.

Suzuki, M. and Smith, J. M. (1975) in *Advances in Chromatography*, Vol. 13, Marcel Dekker, New York.

van Deemter, J. J., Zuiderweg, F. J., and Klinkenberg, A. (1956) *Chem. Eng. Sci.* **5**: 271.

Villermaux, J. (1973) *J. Chromatogr.* **83**: 205.

Vink, H. (1965) *J. Chromatogr.* **20**: 305.

Yamaska, K. and Nakagawa, T. (1973) *J. Chromatogr.* **93**: 1.

Yamaska, K. and Nakagawa, T. (1974) *J. Chromatogr.* **92**: 213.

Yamazaki, H. (1967) *J. Chromatogr.* **27**: 14.

SEDIMENTATION EQUILIBRIUM OF PROTEINS IN DENSITY GRADIENTS

7

JAMES B. IFFT

I. INTRODUCTION

The technique of equilibrium density-gradient ultracentrifugation in the analytical ultracentrifuge was first described by Meselson, Stahl, and Vinograd in 1957 (Meselson *et al.*, 1957). The method provides a thermodynamically rigorous procedure for the determination of the buoyant density, hydration, and solvated molecular weight of polymers having molecular weights larger than about 10,000.

The experiment consists of spinning a dilute solution of the polymer in a concentrated salt solution to equilibrium in an analytical ultracentrifuge, generally the Spinco Model E. The centrifugal field causes a redistribution of the salt molecules in the solution column, thus establishing a density gradient. If the density of the salt solution is initially adjusted so that its density closely approximates that of the protein or nucleic acid, polymer

JAMES B. IFFT, Department of Chemistry, University of Redlands, Redlands, California, 92373. This work was supported in part by a research grant from the National Institute of General Medical Sciences (GM 18871), National Institutes of Health, U.S. Public Health Service.

molecules near the meniscus will be more dense than the solvent there, and the opposite will be true for polymer molecules near the cell bottom. These density differences and the presence of a very large centrifugal force result in the movement of the polymer into a band at a position corresponding to the buoyant density of the polymer. If two polymers are present, they will of course band at different positions in the gradient column provided that they have different densities. Figure 1 schematically displays the banding of two such polymers.

Of the two photographic systems available in the ultracentrifuge, the Schlieren optical system is generally selected for protein studies because the extinction coefficient of proteins is not particularly large, as in the case of nucleic acids. This technique permits the accurate location of band center either through the use of single-sector cells employing an interpolation between the ends of the baseline or double-sector cells with the protein solution in one sector and the solvent in the other. Figures 2 and 3 display the results for a typical protein for these two types of experiments.

Since the discovery of this remarkable technique, most of the applications of it have been in the study of DNA. Meselson and Stahl (1958) showed that the replication of DNA from *Escherichia coli* was semiconservative in their classic experiments with labeled nucleic acids. Doty and co-workers (Sueoka *et al.*, 1959) made the important discovery that the buoyant density of DNA is a linear function of its base composition. This relationship provides a wide variety of separation and identification methods. More

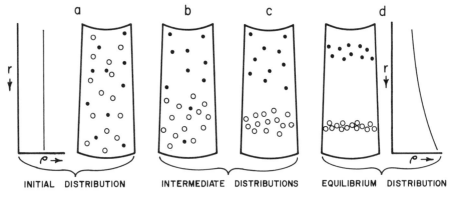

FIGURE 1. Schematic diagram of the principle of sedimentation equilibrium in a density gradient. (a) Initially the particles are distributed uniformly in a solution of uniform density. (b, c) As the gradient forms, the particles seek their respective buoyant densities. The larger particles move more rapidly toward their equilibrium position. (d) At equilibrium, complete resolution has been achieved. The band width is inversely proportional to the mass of the particle (Ifft, 1969).

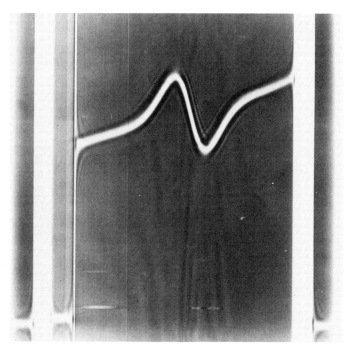

FIGURE 2. Equilibrium Schlieren photograph of 0.1% BMA (bovine serum mercaptalbumin) in CsCl of ρ_e = 1.279 g/ml, pH = 5.08, acetate buffer (μ = 0.01) at 56,100 rev/min and 25°C. Single-sector, 4°, Kel-F centerpiece (Ifft, 1969).

recently, Schmid and Hearst (1969) have provided an unambiguous method of obtaining molecular weights by banding two isotopically labeled DNA samples in the same cell.

The applications in the field of protein chemistry have been considerably more limited (Ifft, 1969, 1973). Ifft and Vinograd (1962, 1966) made an extensive study of bovine serum mercaptalbumin (BMA) in a variety of salt solutions and obtained the buoyant density and hydration of BMA in each solution as well as its correct weight-average molecular weight. Hu *et al.* (1962) showed that proteins can be separated in density gradients in preparative centrifuge tubes. Much of the work in our laboratory in recent years has been devoted to the measurement of the buoyant-titration curves of proteins and polypeptides. The term buoyant titration is understood to mean the measurement of the buoyant density of a polymer as a function of pH. The term was suggested by Professor Vinograd, who reported the buoyant

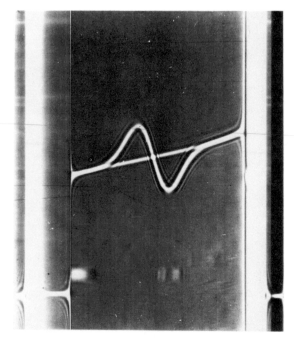

FIGURE 3. Equilibrium Schlieren photograph of 0.1% BMA (bovine serum mer-captalbumin) in 2.59 molal CsCl, pH 5.5, acetate buffer ($\mu = 0.01$) and baseline at 56,100 rev/min and 25°C. Double-sector, capillary-type, synthetic-boundary, 2.5°, filled-Epon centerpiece (Ifft and Vinograd, 1962).

titration of ΦX-174 DNA and some data for poly-L-glutamic acid in 1962 (Vinograd and Hearst, 1962).

II. THEORY

A. Two-Component Theory

The basic relation governing equilibrium in a two-component system can be derived from thermodynamic or kinetic considerations. The most elegant derivation has been provided by Goldberg (1953). The gradient of activity a of the solute with radial position r in the cell is given as

$$\frac{da}{dr} = \frac{M(1 - \bar{v}\rho)\omega^2 r}{RT} \cdot a \tag{1}$$

where M is the anhydrous molecular weight of the solute, $\bar{v}$ is its partial specific volume, ρ is the density of the solution, ω^2 is the square of the angular velocity expressed in rad/sec, and R and T are the usual thermodynamic constants.

This relation has been used (Ifft *et al.*, 1961) to compute density gradients, $d\rho/dr$, from the relation

$$\frac{d\rho}{dr} = \frac{\omega^2 r}{\beta^0} \qquad (2)$$

where β^0 is called the density-gradient proportionality constant. It is a function of the composition of the solution and is strictly valid only at atmospheric pressure as indicated by the superscript zero. It was originally tabulated (Ifft *et al.*, 1961) for five salts and sucrose by means of equation (1) and several graphical procedures. More recently, Ifft *et al.* (1970) have utilized a method described by Trautman (1960) to write a FORTRAN IV computer program to calculate β^0 values for any salt for which adequate thermodynamic data are available. They reported the values shown in Table I for 20 1:1 electrolytes over their entire solubility ranges.

Density-gradient proportionality constants, are needed to determine molecular weights in some instances, to predict resolution, and to determine density distributions in the solution column at equilibrium. The last function is the most important use in protein work. However, it is necessary to know the density at some radial position in the cell. The radial position sought is r_e, the position at which the density of the salt solution equals the uniform solution density prior to centrifugation, ρ_e. The preferred method of computation of r_e at present is the use of the relation (Ifft, 1971)

$$(r'_e)^2 - (r_e)^2 = D(r_b{}^2 - r_a{}^2)^2/48 \qquad (3)$$

where r'_e is the root mean square of the radial positions of the meniscus and cell bottom. D is defined as $g\omega^2/\beta_e{}^2$, where g is the slope of the $\beta(\rho)$ curve at $\rho_e{}^0$ and β_e is the β value at $\rho_e{}^0$. Values of g have been obtained by means of a computer method of least-squares curve fitting and are given in Table II (Almassy *et al.*, 1973).

When the isoconcentration position has been determined, the buoyant density at atmospheric pressure, $\rho_0{}^0$, can be readily determined from the relation

$$\rho_0{}^0 = \rho_e{}^0 + \int_{r_e}^{r_0} (\omega^2 r/\beta^0)\, dr \qquad (4)$$

where the subscript zero denotes properties at band center. If the protein bands sufficiently close to r_e, the integral can be replaced with $(d\rho/dr)_e \cdot (r_0 - r_e)$ A stepwise integration may be required if the band forms near the extremes of the cell.

TABLE I
Variation of $\beta^0 \times 10^{-9}$ with Solution Density
for Several Salts

		$\beta^0 \times 10^{-9}$			
ρ, g/ml	M	MCl	MBr	MI	MNO$_3$
1.05	Cs	7.10	4.70	3.99	4.86
1.075		4.71	3.40	2.81	3.43
1.10		3.61	2.63	2.15	2.45
1.125		2.97	2.16	1.74	1.90
1.15		2.54	1.85	1.47	1.55_0
1.175		2.25	1.63	1.30	1.30_2
1.20		2.04	1.46	1.17	
1.25		1.73_5	1.21_4	0.97_0	
1.30		1.54_5	1.06_0	0.80_5	
1.35		1.41_0	0.96_4	0.71_0	
1.40		1.32_5	0.88_8	0.64_5	
1.45		1.27_0	0.83_8	0.59_0	
1.50		1.22_5	0.80_0	0.55_0	
1.55		1.19_5	0.76_8		
1.60		1.17_0	0.73_0		
1.65		1.15_5	0.69_2		
1.70		1.14_0			
1.75		1.13_0			
1.80		1.12_2			
1.85		1.12_0			
1.90		1.11_8			
1.05	Rb	9.80	7.13	5.50	7.30
1.075		7.56	4.95	3.74	4.70
1.10		5.79	3.75	2.90	3.36
1.125		4.78	3.09	2.35	2.61
1.15		4.15	2.62	2.01	2.09
1.175		3.72	2.34	1.76	1.71
1.20		3.42	2.11_5	1.58	1.41_0
1.25		3.04_0	1.76_5	1.33_5	1.01_5
1.30		2.76_0	1.56_5	1.18_0	0.85_0
1.35		2.46_0	1.43_5	1.06_5	0.73_0
1.40			1.34_0	0.98_5	
1.45			1.27_5	0.93_5	
1.50			1.22_0	0.89_5	
1.55				0.86_0	
1.60				0.83_0	
1.65				$0.80_5{}^a$	
1.05	K	22.1	11.28	7.94	11.0
1.075		16.1_5	7.88	5.50	7.3
1.10		13.1_4	6.18	4.28	5.1

TABLE I (*continued*)

ρ, g/ml	M	\multicolumn{4}{c}{$\beta^0 \times 10^{-9}$}
		MCl	MBr	MI	MNO$_3$
1.125		11.3_8	5.22	3.58	3.9
1.15		10.1_3	4.58	3.11	3.1_3
1.175		$9.0_7{}^a$	4.14	2.78	2.4_6
1.20			3.80	2.55	$1.8_8{}^a$
1.25			3.36	2.20_5	
1.30			3.05	1.95_5	
1.35			2.80^a	1.83_0	
1.40				1.73_0	
1.05	Na	28.7	12.8	8.66	17.1
1.075		22.0	9.51	6.38	11.70
1.10		18.7_0	7.68	5.06	8.88
1.125		16.9_2	6.57	4.30	7.22
1.15		15.8_2	5.88	3.80	6.16
1.175		15.1_1	5.47	3.46	5.38
1.20		$14.2_6{}^a$	5.17	3.19	4.88
1.25			4.49	2.86_0	4.28
1.30				2.81_5	
1.35				$2.94_5{}^a$	
1.05	Li	81.9	15.1	10.54	31.5
1.075		75.8	12.4	8.26	25.6
1.10		75.0	10.7_7	6.86	22.2
1.125		77.2	9.8_6	6.04	20.3_0
1.15		78.8	9.4_9	5.64	18.9_4
1.175		77.5	9.2_7	5.48	17.9_6
1.20		73.3	9.1_4	5.36_5	17.2_6
1.25		57.2	8.9_8	4.92_0	16.2_6
1.30		42.3	8.9_0		15.4_6
1.35			8.9_2		13.7_7
1.40			9.1_4		
1.45			9.5_9		
1.50			10.4_7		
1.55			11.1_1		
1.60			11.3_7		
1.65			11.4_1		
1.70			11.1_2		
1.75			10.6_4		
1.80			10.0_2		
1.85			9.1_5		
1.90			7.6_6		

^{a} Obtained by a short extrapolation.

TABLE II
Slope of $\beta(\rho)$ for CsCl at 25°C

Solution density, g/ml	Slope of $\beta \times 10^{-9}$, $cm^8/g^2\ sec^2$
1.05	−89.718
1.10	−44.987
1.15	−15.475
1.20	−8.045
1.25	−4.929
1.30	−3.225
1.35	−2.207
1.40	−1.392
1.45	−0.996
1.50	−0.752
1.55	−0.540
1.60	−0.414
1.65	−0.311
1.70	−0.233
1.75	−0.172
1.80	−0.109
1.85	−0.040
1.90	−0.041

B. Three-Component Theory

The fundamental relations for noninteracting systems of polymer–salt–water were developed in the original paper by Meselson *et al.* (1957). They found that the distribution of polymer in the band would be Gaussian for a homogeneous sample. The concentration c at a distance δ from band center would be related to the concentration at band center, c_0, by the equation

$$c = c_0 \exp\left(-\delta^2/2\sigma^2\right) \qquad (5)$$

where σ is the standard deviation of the Gaussian band. Their work demonstrated that the molecular weight of the banded species was inversely related to the square of σ:

$$M = \frac{RT}{\bar{v}(d\rho/dr)_{r_0}\omega^2 r_0\sigma^2} \qquad (6)$$

A number of investigators have studied the interaction of the solvent with the polymer and its effect upon equation (6). Williams *et al.* (1958)

were the first to show that a buoyant-density experiment unambiguously yields the net hydration, Γ', of the polymer in grams water/gram polymer:

$$\rho_0 = \frac{1 + \Gamma'}{\bar{v}_3 + \Gamma' \bar{v}_1} \tag{7}$$

This hydration value does not reveal the water required to provide a buoyant composition for the bound salt.

It is apparent from the derivation that this buoyant density must be the density at band center corrected for the effects of pressure. This correction (Hearst et $al.$, 1961) to the buoyant density obtained at atmospheric pressure is made using equation (8).

$$\rho_0 = \rho_0{}^0(1 + \kappa P_{r_0}) \tag{8}$$

where κ is the isothermal compressibility of the solvent of buoyant composition and P_{r_0} is the hydrostatic pressure at the band center (Ifft, 1969). This buoyant density ρ_0 can be substituted into equation (6) as the reciprocal of $\bar{v}$, the partial specific volume of the hydrated species.

The correct density gradient to be used in equation (6) has been the subject of considerable discussion (Baldwin, 1959; Hearst and Vinograd, 1961a,b; Casassa and Eisenberg, 1961; Fujita, 1962; Cohen and Eisenberg, 1968). It is at once apparent that the $composition$ density gradient given by equation (2) and valid at one atmosphere does not represent the total gradient. The effects of pressure are included in the $physical$ density gradient and this effect as well as the compressibility and hydration of the polymer are included in the $buoyancy$ and $effective$ density gradients. The expressions for these four gradients are included in Table III. The quantity $\psi = (\kappa - \kappa_S)/(1 - \alpha)$ can be measured in a series of experiments in which the pressure is varied (Hearst et $al.$, 1961). The quantity α can be determined by banding the polymer in a series of different salt solutions and is computed as

$$\alpha = (\partial \rho_0{}^0 / \partial a_1{}^0)_P (da_1{}^0 / d\rho^0),$$

TABLE III
Four Density Gradients of Significance in Density-Gradient Experiments

Name	Definition
Composition	$(d\rho/dr)_{\text{comp}} = \omega^2 r / \beta^0$
Physical	$(d\rho/dr)_{\text{phys}} = [(1/\beta^0) + \kappa \rho^{02}]\omega^2 r$
Buoyancy	$(d\rho/dr)_{\text{buoy}} = \{1/\beta^0 + [(\kappa - \kappa_S)/(1 - \alpha)]\,\rho^{02}\}\omega^2 r$
Effective	$(d\rho/dr)_{\text{eff}} = \{[(1 - \alpha)/\beta^0] + (\kappa - \kappa_S)\rho^{02}\}\omega^2 r$

where $a_1{}^0$ is the activity of water at band center at atmospheric pressure (Hearst and Vinograd, 1961a). The validity of the effective density gradient in this form has been demonstrated for the protein bovine serum mercaptalbumin (BMA) (Ifft and Vinograd, 1966).

A simpler method of computing $(d\rho/dr)_{eff}$ has been proposed and tested for DNA (Vinograd and Hearst, 1962; Eisenberg, 1967; Schmid and Hearst, 1971; Hearst and Schmid, 1973). ^{14}N- and ^{15}N-labeled samples of the same DNA are banded in a CsCl gradient. The distance between the peaks, Δr, the buoyant density, ρ_0, and the average of the two banding positions, r_0, are measured. These quantities and the change in mass per nucleotide residue upon isotopic substitution, Δm, and the mass of DNA per cesium nucleotide, m, are related to the effective density gradient by equation (9):

$$G = \frac{1 + \Gamma'}{\beta_{eff}} = \frac{\Delta m \rho_0}{m \Delta r \omega^2 r_0} \tag{9}$$

Values of G have been tabulated for Cs_n-DNA (Hearst and Schmid, 1973) and been employed in the correct calculation of apparent molecular weights of dry Cs_n-DNA, $M_{3,app}$, of a number of phage DNAs (Schmid and Hearst, 1969, 1971) utilizing equation (10):

$$1/M_{3,app} = G(\omega^4 r_0{}^2/RT\rho_0)\sigma_{app}^2 \tag{10}$$

The true, anhydrous molecular weight, M_3, is obtained by a linear extrapolation to zero concentration of a plot of $\log_{10} M_{3,app}$ vs. DNA concentration.

It is apparent that equations (6) and (10) are related if we recognize that the M calculated with equation (6) is a solvated molecular weight:

$$M = M_3(1 + \Gamma') \tag{11}$$

and that the correct density gradient to employ with equation (6) is the effective density gradient

$$(d\rho/dr)_{eff} = \omega^2 r_0/\beta_{eff} \tag{12}$$

Unfortunately, values of G have not been obtained for proteins to date. Isotopically labeled proteins are not readily obtained. More seriously, it is rather difficult to band two proteins of average molecular weight in the same gradient column. This is apparent from the standard deviation of the mercaptalbumin bands displayed in Figures 2 and 3.

C. Four-Component Theory

The three-component theory of Schmid and Vinograd (1969) has recently been extended to the case of a protein molecule which selectively binds one of the ions of the solvent to form PX_n or both ions to form PX_nY_m (Sharp et al., 1975). The results of this derivation agree with the equation proposed earlier (Ifft and Vinograd, 1966).

$$\rho_0 = \frac{1 + \sum z_i + \Gamma'_*}{\bar{v}_3 + \sum z_i\bar{v}_i + \Gamma'_*\bar{v}_1} \tag{13}$$

The quantity z_i is the grams of bound ion i per gram of anhydrous, saltfree protein and $\bar{v}_i$ is the partial specific volume of the ion i at that salt concentration. Values of $\bar{v}_{Cs}{}^+$ and $\bar{v}_{Cl}{}^-$ have been computed as a function of concentration (Ifft and Williams, 1967). Γ'_* is the hydration of the protein–salt complex and, as in the case of z_i, it is expressed in units of g water/g anhydrous, saltfree protein. Equation 13 is a logical extension of equation (7) expressing the density of 1 g of protein to which is attached $\sum z_i$ g of salt and Γ'_* g of water as the mass of this unit divided by its volume.

Equation (13) assumes particular significance for our laboratory in that much of our recent work has centered on the measurement of buoyant densities of proteins and polypeptides as a function of pH and the interpretation of ρ_0 in terms of bound ions and water. Relatively little work is being done at present on the application of the technique of density-gradient centrifugation to the determination of the molecular weights of proteins.

D. Resolution

This technique has had some limited application in the field of the separation of proteins. Ifft et al. (1961) defined resolution, Λ, as

$$\Lambda = \Delta r/(\sigma_1 + \sigma_2) \tag{14}$$

where Δr is the difference in radial position of the band centers, and σ_1 and σ_2 are the standard deviations of the two protein bands. Good resolution of two bands is indicated by a large value of Λ which corresponds to a relatively large Δr and small σ values.

Combination of equations (2), (6), and (14) and several reasonable approximations (Ifft et al., 1961) yields

$$\Lambda = \frac{\Delta\rho}{2}\left(\frac{M\beta'}{\rho}\right)^{1/2}_{\bar{r}_0} \tag{15}$$

This derivation assumes that the two solutes differ in buoyant density by $\Delta\rho$ and that they both have molecular weight M. The quantity β' is defined as β/RT and is evaluated at the average of the two banding positions, $\bar{r}_0$.

The resolution is thus seen to be nearly independent of T, with the exception of small variations in activity coefficients with temperature, and is independent of ω. The latter independence is reasonable in that both Δr and σ

are proportional to the reciprocal of the angular acceleration ω. As ω increases, the bands become narrower, but because the density gradient also increases, the band separation decreases and no improvement in resolution is obtained.

One method of improving resolution is apparent from equation (15): Λ is directly proportional to β; thus, salt solutions having the largest density gradient proportionality constants for a given ρ will provide the best resolution. In general, these solutions are provided by the lowest-molecular-weight salts. At a density of 1.3, KBr and RbCl solutions resolve better than CsCl by factors of 1.45 and 1.41, respectively.

There are several ways to interpret the meaning of a given Λ. If $\Lambda = 1$, only one maximum in the concentration curve will be observed. If $\Lambda = 2$, about 5% of the material will be intermixed. For $\Lambda = 3$, virtual resolution is achieved.

Alternatively, an analytical expression has been obtained (Ifft et al., 1961) for the measurement of resolution. If y is defined as the ratio of the sum of the concentrations midway between band centers to the concentration at band center, we obtain

$$y = 2 \exp\left(-\Lambda^2/2\right) \tag{16}$$

for the case of identical initial concentrations and standard deviations.

For proteins of mol. wt. 100,000 g/mole and buoyant density 1.3 g/ml, $\Lambda = 35\Delta\rho$ for CsCl solutions and $50\Delta\rho$ for KBr solutions. Therefore,

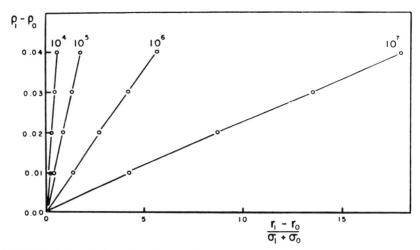

FIGURE 4. Resolution $\Lambda = (r_1 - r_0)/(\sigma_1 + \sigma_0)$ as a function of the difference in buoyant densities of two species. Data apply to RbCl solutions of $\rho_e = 1.33$ g/ml for proteins having the molecular weights indicated on the curves (Hu et al., 1962).

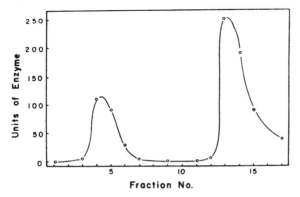

FIGURE 5. Equilibrium distribution of D^2, N^{15} doubly labeled β-galactosidase and unlabeled enzyme in a RbCl gradient in a preparative centrifuge at 500 rev/sec. Density increases with increasing fraction number (Hu *et al.*, 1962).

density differences of 0.058 and 0.40 for CsCl and KBr solutions, respectively, are required to achieve a $\Lambda = 2$. Proteins in this size range could thus be resolved satisfactorily in a density-gradient column in the analytical ultracentrifuge.

Hu *et al.* (1962) have also defined resolution as above and have demonstrated that proteins can be resolved in preparative density-gradient experiments. They presented several plots of practical interest. Figure 4 gives their results for the dependence of $\Delta\rho$ on Λ for proteins of several molecular weights banded in RbCl solutions of $\rho_e = 1.33$ g/ml. The results of the separation of isotopically labeled proteins are given in Figure 5. They grew *E. coli* ML 308 in normal medium and in a doubly labeled ^2D and ^{15}N medium. Extraction of the enzyme β-galactosidase followed by banding in a RbCl gradient in a preparative centrifuge at 500 rev/sec yielded the resolved peaks displayed in Figure 5.

III. EXPERIMENTAL

Most of the procedures for the conduct of density-gradient experiments with proteins have been presented in two earlier works (Ifft, 1969, 1973). A few of the most important and/or more recent changes in these procedures are described below.

A. Solution Preparation

The initial solution should consist of 0.1% protein, 0.01 M buffer, and $\rho_e \simeq \rho_0$. The protein concentration is a nominal value. Concentrations ranging from 0.05 to 0.2% will yield recognizable bands. As is the case with most centrifuge work, the absolute value of protein concentration is not required in the analysis.

Depending on the proximity of the pH of the run to the pK_a of the weak acid, buffer concentrations can range from 0.01 up to perhaps 0.05. The higher concentrations are especially needed at high pH, where CO_2 absorption is especially serious. At the extremes of pH, dilute solutions of HCl or CsOH provide adequate buffering. The effect of pressure on the equilibrium between the weak acid and its salt has recently been investigated (Ifft, 1971). If precise values of pH during centrifugation are required, it is important to select buffers which have a small volume change upon ionization. For some of the amine buffers ΔV is 30 ml/mole, which can lead to pH shifts of 0.1 unit in the center of a CsCl column at full speed.

The salt used to provide the gradient can be selected from any of the numerous compounds for which $\beta(\rho)$ and $n_D{}^{25}(\rho)$ data are available. The $\beta(\rho)$ values for 20 1:1 electrolytes have been computed throughout their entire solubility ranges (Ifft et al., 1970). If β values have not been computed for a salt of interest, a FORTRAN IV program can be obtained from the Chemistry Department of the University of Redlands to facilitate the calculations. The required thermodynamic data are the density as a function of weight percent and activity coefficients as functions of molality.

DNA chemists have been particularly resourceful in securing new and unusual banding media. A recent addition to this list is sodium iothalamate (P. Server, unpublished).

TABLE IV
Density-Refractive Index Relations[a] for Aqueous Salt Solutions at 25°C

$$\rho^{25} = a + b(n_D{}^{25}) + c(n_D{}^{25})^2$$

Salt	Coefficients of equation			Valid density range, g/ml
	a	b	c	
CsCl	1.1584	−10.2219	7.5806	1.00–1.90
CsBr	2.7798	−12.2102	8.1615	1.05–1.50
CsI	4.5245	−13.5833	8.2067	1.05–1.55
RbCl	21.7661	−39.2834	17.7843	1.05–1.35
RbI	−7.1472	4.8363	0.9561	1.05–1.75

[a] All relations are from Ifft et al. (1970).

TABLE V

Density-Refractive Index Relations for Aqueous
Salt Solutions at 25°C

$$\rho^{25} = a(n_D{}^{25}) - b$$

Salt	Coefficients of equation		Valid density range, g/ml
	a	b	
NaCl	4.23061	4.64125	1.00–1.19[a]
KBr	6.4786	7.6431	1.10–1.35[b]
RbBr	9.1750	11.2410	1.15–1.65[b]

[a] Ifft (1969).
[b] Ifft and Vinograd (1966).

The initial solution density is adjusted to approximate ρ_0. If no data are available for a new protein under investigation, a good value to begin with is $\rho_e = 1.30$ g/ml if the protein is to be banded at its pI. The adjustment of the initial density is usually performed by means of measurements of the refractive index and the use of established $n_D{}^{25}(\rho)$ relations. Tables IV and V give quadratic and linear relations for some of the more commonly employed salts.

B. Run Conditions

Unless an integration through the band to obtain a molecular weight will be required, single-sector centerpieces are generally adequate. Kel-F is a satisfactory centerpiece material although it does undergo a permanent, gradual flow after long periods of time at high speed. One-cell runs are an expensive luxury. We routinely run two cells in the An-D or An-H rotors and three or four cells in the An-F rotor. This, of course, requires the use of wedge windows. Figure 6 shows a four-cell run of acetylated BMA which employed $+1°$, $0°$, $-1°$, and $-2°$ wedge windows. Not all multiple-cell runs are this elegant. Occasionally, overlapping menisci or the presence of a precipitate in one cell may obscure the image of another cell in a vital region of the refractive-index-gradient curve. However, we have found multiple-cell runs to be indispensable savers of money and time.

Because almost all data for the salt solutions employed in these experiments are valid only at 25°C, most experiments are conducted at 25°C. Common angular velocities which provide a gradient sufficiently high to form bands of reasonable width range from the top speed of 52,640 rev/min for the An-F aluminum rotor to 67,770 rev/min for the An-H titanium rotor.

FIGURE 6. Equilibrium Schlieren photograph of four cells containing, respectively, from top to bottom: $+1°$, native BMA; $0°$, carbamylated BMA; $-1°$, native BMA: $-2°$, native BMA. $\omega = 52,640$; CsCl; 25°C; single-sector Kel-F centerpieces; An-F rotor (Ellis *et al.*, 1975).

With the exception of poly(Lys) and poly(Orn), which band at very low values of ρ_0 in CsCl and require 40 hr to reach equilibrium, all proteins and polypeptides studied to date have come to equilibrium in 20–24 hr.

C. Data Analysis

A microcomparator is very useful in the measurement of the Schlieren photographic plates. Measurements of both reference edges, the meniscus and cell bottom, can be made with a precision of ± 0.05 mm. The center of a soluble protein band cannot be obtained from the microcomparator image directly. It is apparent from Figure 2 that a rather long interpolation is required. We presently accomplish the location of r_0 by placing a sheet of vellum over the projected image on the microcomparator screen and placing dots at a number of points along the curve and at the inside edge of the outer reference edge. This procedure requires the use of the outer reference edge because the Nikon 6C instrument which we use does not permit the projection of the complete Schlieren frame at one time.

The data from the micrometer heads are converted to real distances from the center of rotation by subtraction of the distance from the outer reference edge, divided by the magnification factor between the plate and the cell, from the radial position of the outer reference edge. The same procedure is followed for the tracing except that the magnification factor used must be between the tracing and the cell.

Alternatively, a photographic enlarger can be used to trace the entire image and all measurements obtained directly from the tracing (Ifft, 1969).

D. Computation

A FORTRAN IV program is available from the Department of Chemistry, University of Redlands, to carry out all of the necessary computations required to obtain a ρ_0 value from the corrected value of n_D^{25}, ω in rev/min, the slope g of the $\beta(\rho)$ plot at ρ_e, β at ρ_e, the isothermal compressibility coefficient κ of the salt solution at ρ_0, and the five pertinent radial positions in the cell.

A simpler program has been written by Dr. Ib Svendsen, who was a visitor in our laboratory from the Carlsberg Laboratorium in 1973–74. This program is written in BASIC language and is run on a computer terminal connected to a Hewlett-Packard 2000 B computer. Input data are the values of ρ_e and ω^2 in (rad/sec)^2 obtained from tables, g and β at ρ_e, and the five pertinent radial positions in the cell. As above, output data are ρ_0^0 and ρ_0. This program is reproduced below (Table VI).

E. Use of a Density Marker in the Determination of ρ_0

It has been apparent to us for some time that the measurement of the buoyant densities of proteins would be simplified if a density marker were available. Greater precision than the ± 0.001 g/ml available from refractometric measurements should be possible and the method would be independent of changes in the salt concentration during the run caused by evaporation. This method has been employed in the measurement of ρ_0 values for DNA for

TABLE VI
BASIC Program for the Computation of Buoyant Densities

```
     LIST
     BOYDEN
10   PRINT TAB(12); "CALCULATION OF BUOYANT DENSITY IN A SALT GRADIENT"
20   PRINT
30   PRINT "DATA SHOULD BE ENTERED IN LINES 290 TO 320. AT THE END OF"
40   PRINT "CALCULATIONS DELETE THE DATA BY TYPING: DEL-290,320"
50   PRINT
51   READ A
52   IF A = −1 THEN 325
53   READ B,C,D,E,F,G,H,X,Y
60   PRINT "A = REFDS"A; TAB(31); "F = SLOPE" F
61   PRINT "B = DSUBA"B; TAB(31); "G = OMEGA − SQ" G
62   PRINT "C = DSUBB" C; TAB(31); "H = RHO(E)"H
63   PRINT "D = DSUBO"D; TAB(31); "X = RUN NR"X
70   PRINT "E = BETA"E; TAB(31); "Y = FINAL PH"Y
80   PRINT
90   PRINT TAB(14); "RUN NUMBER"; TAB(30); "PH"; TAB(45); "RHO(0)"
100  PRINT
140  LET I = A/1.595
150  LET J = 7.325 − B/I
160  LET K = 7.325 − C/I
170  LET L = 7.325 − D/I
180  LET M = SQR((J^2 + K^2)/2)
190  LET N = SQR((L^2 + J^2)/2)
200  LET O = SQR(M^2 + F*G*.01*((K^2 − J^2)/(E^2*48))
210  LET P = H + (L − O)*G*O/(100*E)
220  LET Q = H + (N − O)*G*O/(100*E)
230  LET R = 10*Q*G*(L^2 − J^2)/2.026
240  LET S = P*(1 + .000001*35*R)
250  PRINT TAB(15);X; TAB(28); Y; TAB(43); S
260  PRINT
270  PRINT
280  GO TO 51
290  DATA 34.785,27.171,2.688,15.37,1.567,3.44,3.0387,1.291,543,5.616
321  DATA −1
325  END
     RUN
     BOYDEN
```

```
          CALCULATION OF BUOYANT DENSITY IN A SALT GRADIENT
     DATA SHOULD BE ENTERED IN LINES 290 TO 320. AT THE END OF
     CALCULATIONS DELETE THE DATA BY TYPING: DEL − 290,320
     A = REFDS 34.785                 F = SLOPE 3.44
     B = DSUBA 27.171                 G = OMEGA − SQ 3.0387
     C = DSUBB 2.688                  H = RHO(E) 1.291
     D = DSUBO 15.37                  X = RUN NR 543
     E = BETA 1.567                   Y = FINAL PH 5.616
               RUN NUMBER        PH          RHO(0)
                   543          5.616        1.28919
     DONE
```

some time (Schildkraut *et al.*, 1962). The DNA under investigation is banded in the same gradient column with a marker DNA. The most frequently used standard is *E. coli* DNA, which has been assigned a value of $\rho_0{}^0 = 1.7100$ g/ml. The $\rho_0{}^0$ value of the unknown DNA is obtained from the following relations:

$$\rho_2 = \rho_1 + \int_{r_1}^{r_2} (\omega^2 r / \beta_B) \, dr \qquad (17)$$

For small values of $(r_2 - r_1)$, and especially in the case of DNA, the density-gradient proportionality constant, β_B, for the buoyancy density gradient (see Table III) can be considered a constant. The radial distances r_1 and r_2 represent the band centers of the standard and unknown DNAs and ρ_1 and ρ_2 their respective buoyant densities. Thus, equation (17) can be directly integrated to obtain the density difference:

$$\rho_2 - \rho_1 = (\omega^2 / 2\beta_B)(r_2{}^2 - r_1{}^2) \qquad (18)$$

Values of β_B have been published (Hearst and Schmid, 1973) for DNA for three salts. These values cannot be used for proteins; they are valid only for native DNA which bands near a $\rho_0{}^0$ value of 1.710 g/ml. Fortunately for the DNA chemist, the $\beta(\rho)$ plot for the compositional density gradient (see Table III) has an almost zero slope at $\rho \simeq 1.7$. This is not the case for $\rho \simeq 1.3$, where most proteins band. If a marker can be obtained for protein work compositional β values will probably have to be used because of the diversity of prosthetic groups present and differences in hydration and ion binding. Both this problem and that of the steep slope of the $\beta(\rho)$ plot at $\rho \simeq 1.3$ can be avoided by banding the unknown protein and the marker close together.

Because of the large σ values of proteins described above and the necessity of banding the protein and the marker with a small $(r_2 - r_1)$, it occurred to us to explore the use of density-gradient marker beads provided by Reproductive Systems, Inc. This company markets gradient beads in densities differing by 0.1 g/ml, from $\rho = 1.1$ to $\rho = 1.9$ g/ml. Unfortunately, there is a rather considerable variation in density in a given set of beads and they must be individually calibrated. Initial efforts to calibrate the beads in a CsCl gradient were abandoned because the calibration still depended on the initial $n_D{}^{25}$ measurements and because it was observed that the beads began to physically deteriorate after two or three runs in CsCl.

The technique of Lang (Linderstrøm-Lang and Lanz, 1938; Linderstrøm-Lang *et al.*, 1938; Hvidt *et al.*, 1954) was employed to perform this calibration at ambient pressure. The 1G density-gradient column described in the Carlsberg papers was constructed, filled with bromobenzene and benzene, and carefully stirred. The column was calibrated with four drops

TABLE VII
Calibration and Determination of Linearity of a *1g* Bromobenzene–
Benzene Density Gradient in a Carlsberg Column[a]

Drop	$\rho(n_D{}^{25})$	$\Delta\rho(n,n+1)$	$\Delta x(n,n+1)$	$\Delta\rho/\Delta x$
1	1.239			
		0.029	5.602	0.0051_8
2	1.268			
		0.034	6.540	0.0052_0
3	1.302			
		0.024	4.625	0.0051_9
4	1.326			

[a] Data are from R. Preston and Ifft (unpublished).

FIGURE 7. Equilibrium Schlieren photograph of a density marker bead banded in a CsCl gradient of $\rho_e = 1.33$ g/ml at 56,100 rev/min and 25°C. Single-sector, 4°, Kel-F centerpiece (R. Preston and Ifft, unpublished).

of CsCl solution whose densities were established refractometrically. The droplet and marker banding positions were carefully measured with a traveling microscope. The data obtained (R. Preston and Ifft, unpublished) are presented in Table VII and indicate that the column is quite linear and that precise marker ρ values can be obtained with it.

After calibration in the $1G$ gradient column, the beads were carefully marked and inserted one at a time into a centrifuge cell before cell assembly and run in a regular banding experiment in CsCl. Figure 7 shows how the image of the marker bead is displayed in a CsCl gradient. The buoyant densities of BMA calculated from the $n_D{}^{25}$ data and the calibrated marker density were identical within ± 0.001 g/ml, the limit of the refractometric accuracy.

The use of these marker beads continues under development in our laboratory. Further problems requiring study are the pressure dependence of the bead densities, the use of KCl solutions of more precisely known density, either from data in the International Critical Tables or measurement with the new Anton Paar densimeter, to calibrate the column, and the possibility of solute binding to the marker beads.

IV. RESULTS

A. Proteins at the Isoelectric Point

The primary data which can be obtained from a buoyant-density experiment are the values of ρ_0 and σ. As discussed earlier, these data can be employed under all conditions to compute Γ' and under some conditions to calculate M and Γ'_*.

Cox and Schumaker (1961) were the first to report the results of banding proteins in salt density gradients. They selected the precipitating solvent CsCl, $(NH_4)_2SO_4$ and used absorption optics to record the narrow band positions. Buoyant densities $\rho_0{}^0$ were measured and Γ' values computed for seven proteins. They also computed hydration values from the relation

$$M_h = M_a + \alpha M_w \tag{19}$$

where M_h, the hydrated molecular weight, is calculated from the observed standard deviation, M_a is the anhydrous molecular weight, and α is the hydration expressed in moles water/mole protein. This procedure is subject to considerable uncertainty in that effective density gradients were not employed to calculate M_h and because this molecular weight includes bound salt as well as bound water.

Ifft and Vinograd (1962, 1966) reported in two papers their studies of BMA. Schlieren optics were used to record soluble bands, in CsCl and several other salt gradients. Figures 8 and 9 display the results of one run and demonstrate that the distribution of protein in a CsCl gradient is Gaussian, as predicted by equation (5). The deviations from ideal Gaussian behavior are largely due to the nonconstant density gradient. Techniques were developed for the simultaneous recording of an accurate baseline, and a number of the assumptions and approximations inherent in the original derivations were examined. BMA was banded in six different salt solutions (CsCl, CsBr, CsI, Cs_2SO_4, KBr, and RbBr). The buoyancy density gradient was computed from experiments in which BMA was banded in two cells in the same experiment in salt solutions of slightly different initial densities, $\rho_{e,2}^0$ and $\rho_{e,1}^0$, and the buoyancy gradient computed as

$$\left(\frac{d\rho}{dr}\right)_{\text{buoy}} = \frac{\rho_{e,2}^0 - \rho_{e,1}^0}{r_{0,1} - r_{0,2}} \tag{20}$$

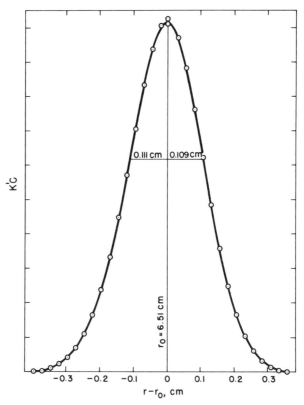

FIGURE 8. Equilibrium distribution of the concentration of BMA in a CsCl gradient. Obtained by numerically integrating the areas under the protein gradient curve in Figure 3 (Ifft and Vinograd, 1962).

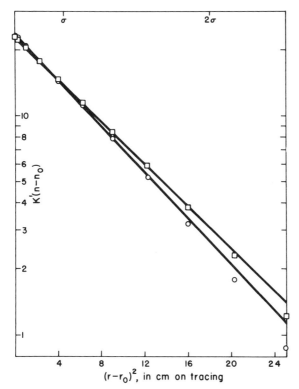

FIGURE 9. Logarithmic plot of the concentration curve derived in Figure 8. ($\bigcirc$) Outer half of distribution; ($\square$) inner half of distribution (Ifft and Vinograd, 1962).

where $r_{0,1}$ and $r_{0,2}$ are the radial positions of the band centers in the two cells. From these data and data available in the literature or obtained in these experiments which yielded σ values, the effective density gradient (Table III) was computed. This density gradient was used in equation (6) to compute molecular weights, which include hydration and salt binding, in each salt solution. Independent measurements of salt binding were made. The partial specific volume of BMA, $\bar{v}_3$, was measured pycnometrically (Ifft and Vinograd, 1962) in a CsCl solution of buoyant composition. It was found to be 0.736 ml/g, not significantly different from the accepted value in water of 0.734 ml/g. The total hydration of the protein–salt complex, Γ'_* [equation (13)], could then be computed. Finally, an anhydrous molecular weight was obtained which agreed with the two-component sedimentation equilibrium value.

The buoyant density of a polymer is an intrinsic property of each molecule in the same way that molecular weights, partial specific volumes, and specific rotations are. They reflect to a first approximation the monomer composition of the polymer. This property has been exploited rather fully by DNA chemists in determining G-C contents, as indicated by the over 300 entries for the buoyant densities of DNAs in the current edition of the *Handbook of Biochemistry* (Mandel, 1972). Because no such extensive data have been compiled for proteins, we have begun the measurement of the buoyant densities of a wide variety of proteins. Table VIII presents the current status of this investigation (Thorpe and Ifft, unpublished). The pH values are within 0.1 unit of the pI value for each protein. Net hydrations, Γ', were computed from published $\bar{v}$ values and the measured buoyant densities with the Williams *et al.* (1958) relation [equation (7)]. It can be seen that the ρ_0 values for all proteins in the soluble form at their isoelectric pHs which

TABLE VIII

Buoyant Densities and Hydrations of Proteins in CsCl Solutions
at the Isoelectric Point at 25°C

Protein	pH	ρ_0 g/ml	Γ' g H_2O/g protein
Hemoglobin (human)	7.04	1.250	0.26
Myoglobin (whale)	6.72	1.256	0.27
Hemoglobin (horse)	6.55	1.264	0.20
Hemoglobin (pig)	6.58	1.272	0.17
β-Lactoglobulin	5.39	1.272	0.16
Aldolase (rabbit muscle)	6.56	1.273	0.20
Hemoglobin (dog)	—	1.274	0.17
BMA	5.50	1.282	0.20
Histone (calf thymus)	9.12	1.284	0.18
Ovalbumin	4.71	1.284	0.14
Cytochrome c (horse heart)	7.91	1.289	0.31
Chymotrypsinogen A	9.45	1.293	0.23
Catalase	5.48	1.299	0.17
7 S IgG	7.30	1.300	0.13
Conalbumin	6.28	1.306	0.14
Ribonuclease	10.55	1.306	0.25
Concanavalin A	8.07	1.308	0.15
α-Chymotrypsin	8.65	1.310	0.12
Trypsinogen	9.11	1.312	0.11
Lysozyme (chicken)	10.50	1.319	0.23
Carboxypeptidase	6.72	1.319	0.16
Pepsin	0.76	1.319	0.035
Edestin	5.62	1.328	0.037
Pepsinogen	3.79	1.332	0.003
Hemocyanin	5.17	1.339	0.030
Gliadin	5.91	1.366	0.030

have been measured to date fall within the range 1.25–1.37 g/ml. All hydration values are in the range 0.03–0.31 g H_2O/g protein. These numbers are consistent with hydration values obtained by other techniques for proteins (Hade and Tanford, 1967). Kuntz and Kauzmann (1974) have recently reviewed our hydration data and discussed a variety of methods which are available for the measurement of Γ'.

We are currently seeking correlation coefficients between the measured buoyant densities and the amino acid compositions. Residues have been grouped as ionic, polar, nonpolar, acidic, and basic. The best correlation observed to date is for plots of buoyant density vs. the weight fractions of polar and nonpolar residues. Figure 10 displays these data. It is apparent that although definite trends appear, variations in prosthetic groups,

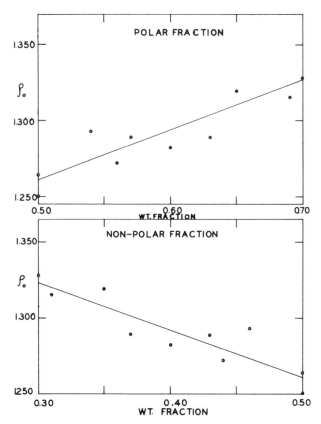

FIGURE 10. The buoyant densities of proteins in CsCl at 25°C as a function of the weight fraction of polar and nonpolar residues (Thorpe and Ifft, unpublished observations).

differences in ion binding or hydration, or other factors are sufficient to preclude any direct linear relationships as in the case of DNA.

B. Buoyant Titrations of Proteins

We have measured the complete buoyant titrations of three proteins and partial curves for three other proteins to date. These experiments consist of 10–15 buoyant density experiments in CsCl solutions with 0.01–0.03 M buffers to yield the desired pHs.

1. Results

The six buoyant titration curves for native, soluble proteins which have been obtained to date are presented in Figures 11–14.

Williams and Ifft (1969) reported the first buoyant titration for a protein (Figure 11). Buoyant densities were measured for BMA for pH 5–12.5. In the figure, uncertainty in ρ_0 in each measurement is reflected by the size of the circles.

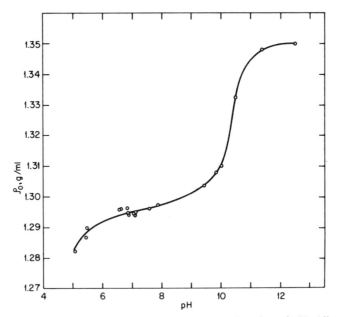

FIGURE 11. Buoyant density, ρ_0, of BMA in CsCl as a function of pH. All runs performed in 4°, single-sector, Kel-F centerpieces at 56,100 rev/min and 25°C. Diameters of circles represent maximum experimental uncertainties (Williams and Ifft, 1969).

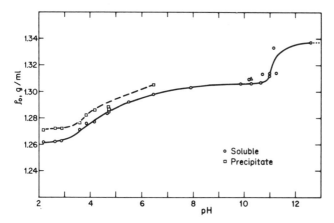

FIGURE 12. Buoyant densities of soluble and precipitated ovalbumin as a function of pH; CsCl; 25°C (Ifft, 1971).

The work on ovalbumin (Ifft, 1971) extended the buoyant titration of a protein to a pH at which all of the carboxyl groups were protonated. Figure 12 displays these data for the protein in both the soluble and precipitated forms. It has often been observed at low pH that both soluble and precipitate forms are found in the same solution and display their characteristic bands. Figure 15 is an example of this behavior for ovalbumin. The light band just to the right of band center of the soluble material represents light lost from the optical system due to the large gradient produced by a narrow band of aggregated protein.

The data in Figure 13 are for a pooled sample of human immuno-gamma-globulin (IgG) (Ruark and Ifft, 1975). The data cover the pH range of stability of IgG. Partial titration curves for conalbumin, cytochrome c,

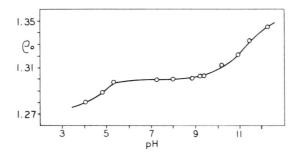

FIGURE 13. Buoyant density of human IgG as a function of pH; CsCl; 25°C (Ruark and Ifft, 1975).

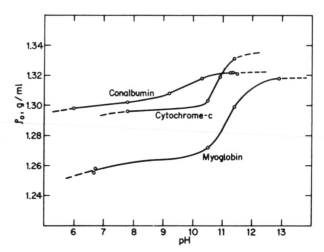

FIGURE 14. Buoyant densities of conalbumin, cytochrome *c*, and myoglobin as a function of pH; CsCl; 25°C (J. S. V. Zil and Ifft, unpublished observations).

and myoglobin are given in Figure 14 (J. S. V. Zil and Ifft, unpublished observations).

2. Interpretation

The most complete interpretation has been done for BMA. The reason for this is that ion-binding data are available at the pI (Scatchard *et al.*, 1957). BMA binds 53 chloride ions at the pI and hence 53 cesium ions must be associated with the buoyant species to provide charge neutrality. Equation (13) was employed to compute Γ'_* and a value of 0.48 g H_2O per gram of protein was obtained. Similarly, the composition of the buoyant protein at pH 12.5 was assumed to be the protein plus 100 bound cesium ions, the number of carboxylate groups in the protein. Again a hydration value could be computed, and Γ'_* was found to be 0.45 g H_2O per gram of protein. Because the decrease in hydration was so slight over the pH interval of 5.4–12.5, Γ'_* was assumed to vary linearly over this range. Thus, the values of ρ_0, $\bar{v}_3$, $\bar{V}_{Cs^+}$, $\bar{V}_{Cl^-}$, and Γ'_* were available at all pH values over this interval for use in equation (21):

$$\rho_0 = \frac{1 + \nu_{Cs^+}(M_{Cs^+}/M_3) + \nu_{Cl^-}(M_{Cl^-}/M_3) + \Gamma'_*}{\bar{v}_3 + \nu_{Cs^+}(M_{Cs^+}/M_3)(\bar{V}_{Cs^+}/M_{Cs^+}) + \nu_{Cl^-}(M_{Cl^-}/M_3)(\bar{V}_{Cl^-}/M_{Cl^-}) + \Gamma'_*\bar{v}_1}$$

$$(21)$$

Equation (21) could only be solved if a second relation involving ν_{Cs^+} and ν_{Cl^-} were obtained. The conservation of charge equation provides

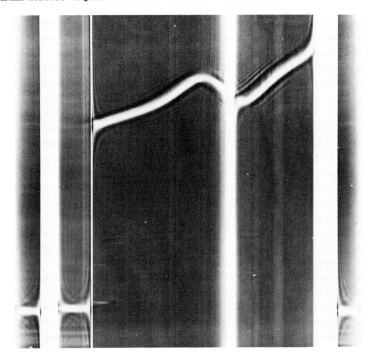

FIGURE 15. Equilibrium Schlieren photograph of 0.1% ovalbumin in CsCl of $\rho_e =$ 1.278 g/ml, pH = 4.66, acetate buffer ($\mu = 0.01$) at 56,100 rev/min and 25°C. Single-sector, 4°, Kel-F centerpiece (J. S. V. Zil and Ifft, unpublished observations).

this relationship. If ϕ represents the algebraic charge on the protein due to the state of ionization of the residues, then

$$\nu_{Cs^+} = \nu_{Cl^-} - \phi \qquad (22)$$

Equations (21) and (22) were combined to yield ν_{Cs^+} and ν_{Cl^-} as a function of pH. Figure 16 presents these data. The changes in the numbers of bound ions over pH intervals when specific groups were titrating were used to calculate classes of binding sites and the numbers of ions bound to specific residues.

Similar computations cannot be performed for the other titration curves because no data are available at any pH for ion binding. The following more direct observations were made for these curves.

The general shapes of all curves obtained to date are the same. The curves of the proteins for which low-pH data are available demonstrate a modest inflection point at about pH 4.5. This is consistent with the pK_{int} of the

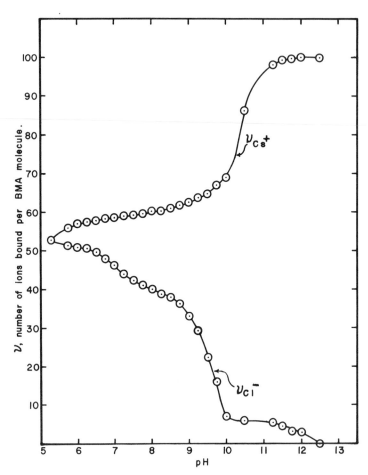

FIGURE 16. The number of cesium and chloride ions bound per molecule of BMA as a function of pH; CsCl; 25°C (Williams and Ifft, 1969).

aspartic and glutamic residues of proteins. All ρ_0–pH plots are rather flat from pH 6 to pH 9. This region corresponds to the titration of the histidine residues which are generally present in relatively small numbers in these proteins. All six proteins display large inflections at high pH. This corresponds to titration of the lysine and tyrosine residues accompanied by loss of heavily hydrated Cl^- ions or gain of Cs^+ ions which are quite dense [ρ_{Cs^+} at $\rho_{soln} = 1.3$ is 6.7 g/ml (Ifft and Williams, 1967)]. This high-pH inflection occurs at pH $= 10.8 \pm 0.3$ for four of the proteins. The value of 9.5 for conalbumin indicates that this protein contains abnormal lysines and/or tyrosines.

Another correlation which has been made is to relate the mole fraction of ionizable lysines and tyrosines present to the density changes observed at pH 8.5–12.5. A good correlation was found for three of the five proteins, with a straight line passing through the origin. These calculations are complicated by the lack of data on the numbers of lysines and tyrosines which titrate with normal pK values.

Finally, hydration values, Γ', can be computed for each protein at pI if published $\bar{v}_3$ data are available (see Table VIII). Net hydrations of the salt-free protein can be computed at other pH values from measured buoyant densities, but it requires the assumption that $\bar{v}_3$ of the protein does not change with pH.

Values of Γ'_* have been computed for ovalbumin at the two extremes of pH where the ionization state of the molecule is known and reasonable assumptions made regarding ion binding. Values of 0.23 and 0.41 g H_2O per gram of protein were obtained at pH 2 and 12.5, respectively. This behavior is in contrast to that observed for BMA.

C. Buoyant Titrations of Synthetic Polypeptides

Because of the difficulties encountered in interpreting the buoyant titrations of proteins, studies of synthetic homo- and copolypeptides have been made. These model compounds have provided information as to how individual residues affect the buoyant densities of proteins. The measurement of ρ_0 values when no ionization occurs, the magnitude of $\Delta\rho$ upon ionization, and the inflection pH values have increased our understanding of the effect of individual residues in proteins on buoyant densities.

1. Buoyant Titrations of Ionizable Homopolypeptides in CsCl

Buoyant titrations in CsCl of six ionizable homopolypeptides have been measured (Almassy et al., 1973). The results are presented in Figure 17. The ρ_0 values of all polypeptides increase with increasing pH. In all cases, an inflection pH occurs near the pK of the monomer amino acid. The buoyant density is independent of pH above and below this marked increase. In the case of the glutamic acid and tyrosine residues, which become negatively charged with increasing pH, the density increases because of the binding of Cs^+ ions whose hydrated densities are greater than that of the neutral residue. The density of the basic residues increases because hydrated chloride ions of density less than the neutral residues are lost.

As indicated in Figure 17, all homopolypeptides are soluble in the buoyant CsCl solutions when present in the charged state. All neutral homopolypeptides are banded as precipitates.

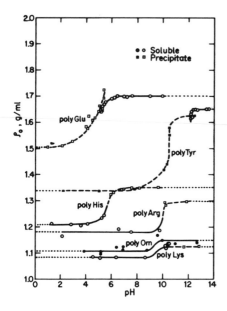

FIGURE 17. Buoyant densities of six ionizable homopolypeptides as a function of pH; CsCl; 25°C (Almassy *et al.*, 1973).

Table IX tabulates the actual ρ_0 values at low and high pH for each homopolypeptide and also gives the observed density increment between these two values. The densities range between a low value of 1.084 g/ml for poly(Lys) at low pH to 1.700 g/ml for poly(Glu) at high pH. This is somewhat surprising in view of the relatively narrow density range of 1.25–1.37 g/ml for proteins discussed above. The lowest buoyant densities correspond to the three residues, lysine, ornithine, and arginine, which have side chains of four or five carbon and nitrogen atoms ending in an ionizable nitrogen group.

TABLE IX
**Buoyant Densities of Homopolypeptides in CsCl
at 25°C**

Homopolypeptide	Buoyant densities, ρ_0		
	Low pH	High pH	$\Delta\rho_0$
poly (Lys)	1.084	1.122	0.038
poly (Orn)	1.111	1.134	0.023
poly (Arg)	1.177	1.297	0.120
poly (His)	1.210	1.350	0.140
poly (Tyr)	1.339	1.651	0.312
poly (Glu)	1.505	1.700	0.195

The cyclic residues, histidine and tyrosine, have intermediate densities. The glutamic residues have the highest density and the shortest side chain.

Figure 17 visually demonstrates the existence of two classes of residues: those which titrate at about pH 5 and the groups which are deprotonated at a pH of about 10. The abnormally low inflection pH of poly(Arg) will be discussed later.

Preferential hydrations, Γ', of the neutral homopolypeptides were calculated from equation (7). The data of Table IX provide the buoyant densities, $\bar{v}_1$ again is assumed to be 1.00 ml/g, and values of $\bar{v}_3$ were computed using the methods and data of Cohn and Edsall (1943). A wide range of hydrations is observed in Table X. The Γ' values, as discussed earlier, represent minimum hydration values and do not reveal additional water which may be bound to the polymer in order to provide a buoyant composition for any bound salt.

Γ'_* values also are given in Table X. They were computed using the above $\bar{v}_3$ values and the assignment of one Cs^+ to each carboxylate group and one Cl^- to each ammonium group. In all cases, the hydrations of the homopolypeptide–ion complex are several times larger than that of the neutral homopolypeptide.

The buoyant titration of poly(Orn) was measured to determine the effect of one less methylene group in comparison with poly(Lys). The assumptions of additive volumes and a density of 0.859 g/ml for the methylene group permitted the accurate prediction of the density of poly(Orn) at both low and high pH from the data for poly(Lys). The success of this calculation suggests that satisfactory predictions of buoyant densities of other homologs, such as poly(Asp) from the data for poly(Glu), will be possible.

These data for ionizable homopolypeptides have been used successfully to predict the low-pH density increments of two proteins. As the pH is raised from 3 to 6, the carboxyl groups titrate. If their behavior in the protein is

TABLE X

Preferential Hydrations of Neutral Homopolypeptides, Γ', and
Homopolypeptide–Ion Complexes, Γ'_*, in CsCl at 25°C

Homopolypeptide	$\bar{v}_{3'}$ ml/g	Γ', g H_2O/g polypeptide	Γ'_*, g H_2O/g polypeptide
poly (Glu)	0.661	0.01	0.91
poly (Tyr)	0.712	0.14	0.66
poly (His)	0.665	0.29	1.16
poly (Arg)	0.694	0.34	1.31
poly (Orn)	0.777	0.89	1.99
poly (Lys)	0.819	0.70	2.31

similar to that in poly(Glu), the $\Delta\rho$ for the protein can be obtained as the density increment for poly(Glu) (0.195 g/ml) times the mole fraction of carboxylic groups in the protein. These calculations yield predictions of 0.026 and 0.023 g/ml for ovalbumin and IgG, respectively. The actual values of 0.033 g/ml and 0.021 are in good agreement and indicate that most of the carboxyl residues in the proteins do titrate at normal pK values and that they are free to interact with the solvent in approximately the same manner as in the homopolypeptide. It is apparent that this homopolypeptide data will be useful in interpreting and predicting the behavior of proteins in buoyant density gradients.

2. Buoyant Titrations of Nonionizable Homopolypeptides in CsCl.

These polymers are insoluble in CsCl and for this reason have not been studied as extensively as the ionizable residues which are all soluble through-out at least a portion of the pH range. The results for poly(Gly) and poly(Ala) are given in Figure 18. They are most surprising. Although there are no groups in either polymer which dissociated a proton, it is apparent that both polymers display an increase in ρ_0 with increasing pH. The inflections are of the order of magnitude of the poly(Lys) and poly(Orn) values and occur at about pH 7–8. No satisfactory explanation of this observation is available at present.

3. Buoyant Titrations of Copolypeptides in CsCl

Studies have been made (Sharp et al., 1975) of the buoyant behavior of polypeptides synthesized from two amino acids. Although the condensa-

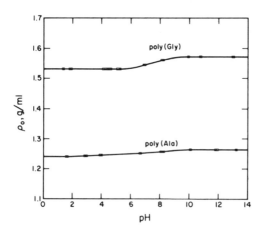

FIGURE 18. Buoyant densities of two nonionizable homopoly-peptides as a function of pH; CsCl; 25°C (Sharp et al., 1975).

tion yielded polypeptides of unknown sequence, the amino acid compositions were accurately measured. These compounds have served as models in our use of the homopolypeptide data given above to predict the buoyant properties of proteins.

The buoyant titrations of five ionizable copolypeptides, poly(Glu$^{60.9}$ Lys$^{39.1}$), poly(Glu$^{90.9}$Ala$^{9.1}$), poly(Glu$^{93.8}$Tyr$^{6.2}$), poly(Glu$^{54.5}$Tyr$^{45.5}$), and poly(Lys$^{51.4}$Tyr$^{48.6}$), were measured in CsCl solutions. As expected, the buoyant densities at all pH values of the copolymers are intermediate between the ρ_0 values of the corresponding homopolypeptides.

Figure 19 displays the data for the glutamic acid–lysine copolymer. The measured buoyant densities displayed there are rather typical of the data for the other four copolypeptides. Inflections are noted at approximately the same pH as was observed for the homopolypeptides.

Several different approaches have been employed to provide the best predictive method to accurately reproduce the measured buoyant titration curve from the homopolypeptide data. The additive volume relationship, equation (23), is physically reasonable and can be derived readily:

$$\frac{1}{\rho_p} = \frac{w_a}{\rho_a} + \frac{w_b}{\rho_b} \tag{23}$$

where ρ_p is the predicted buoyant density, ρ_a and ρ_b are the measured buoyant densities of homopolypeptides a and b at a given pH, and w_a and w_b are the weight fractions of the two homopolypeptides.

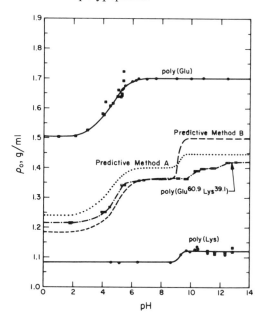

FIGURE 19. Measured and predicted buoyant densities of poly(Glu$^{60.9}$Lys$^{39.1}$) as a function of pH; CsCl; 25°C (Sharp et al., 1975).

Two methods of calculation have been applied to equation (23). In both methods, one ion of the opposite charge is assumed to be bound to the ionized residue. The methods differ in the treatment of the hydration calculation. In Method A, the hydration of the residues is assumed to be independent of the state of ionization, and Γ' values computed from equation (7) for each homopolypeptide are used throughout the entire pH range. In Method B, the hydration of the neutral residues is computed as above, but Γ'_* values are employed for the charged residues [equation (13)]. Method B is complicated by the necessity of employing an iterative procedure to account for the difference in partial specific volume of the bound ion at the density of the homo- and copolypeptides. Figure 19 indicates that Method A is somewhat superior to Method B. This was generally found to be true for the other copolypeptides as well. It is apparent from the data obtained thus far, however, that neither of these methods provide predicted buoyant densities with great accuracy. Values of $\Delta\rho$ of several hundredths of a density unit are common in many regions of the curves.

4. Buoyant Titrations of Homopolypeptides in Several Salt Solutions

The effect of different salt solutions on the buoyant density of a protein has been known for some time (Ifft and Vinograd, 1966). The buoyant densities of BMA in CsCl, CsBr, and CsI are 1.282, 1.315, and 1.347, respectively. Thus, it is of interest to observe the effect of various salts on several ionizable homopolypeptides. An anion series would be especially interesting for the basic residues and a cation series for the acidic residues in view of the model of one-to-one binding. Because of the limited solubility ranges of some of the salts and the high buoyant densities of poly(Glu) and poly(Tyr), studies have been limited to poly(His) and poly(Lys).

Figure 20 displays the results obtained for poly(Lys) in CsCl, CsBr, and CsI. (N. Fujita, K. Kinzie, and Ifft, unpublished observations). As in the case of BMA, the low-pH buoyant densities increase with increasing mass of the anion—1.085, 1.221, and 1.458. Although partial specific volumes of the Br^- and I^- ions are not available for these concentrated salt solutions, it is apparent that the larger mass of these ions or the decreased water activity of the buoyant solutions have a profound effect on ρ_0 values.

The buoyant titration curves in CsBr and CsI are interesting in that they represent the first cases in which the buoyant density of a polypeptide decreases with increasing pH. As the residues titrate, the leaving Br^- and I^- ions are less heavily hydrated than the Cl^- ions, thus causing the decrease in ρ_0.

These curves also demonstrate that the buoyant densities of the neutral homopolypeptide at high pH converge toward the same value. The curves cannot be extended much further because of the complication involved in

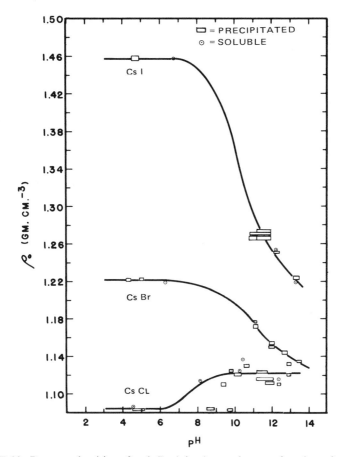

FIGURE 20. Buoyant densities of poly(Lys) in three salts as a function of pH 25°C (N. Fujita, K. Kinzie, and Ifft, unpublished observations).

computing the CsOH–CsCl gradients and because of hydrolysis of the polypeptide. The convergence of these curves indicates that water activity is not a dominant factor in determining the ρ_0 of the neutral polymer.

Table IX gives the data obtained for these two basic polymers at low and high pH. It is apparent that the nature of the anion has a profound effect on the buoyant density, the nature of the cation has little effect, and the buoyant density of the high-pH neutral form of the polymer is almost independent of the banding medium.

D. Buoyant Titrations of Chemically Modified Proteins

The contribution of each residue to the hydration and ion binding, which determine in part the buoyant density of a protein, is needed in order

TABLE XI

Buoyant Densities at Low and High pH of poly (Lys)
and poly (His) in Various Salt Solutions at 25°C

| | | Buoyant densities | |
	Solvent	Low pH	High pH
poly (Lys)	CsCl	1.085	1.120
	CsBr	1.221	1.134
	CsI	1.458	1.224
	RbCl	1.109	1.112
	RbBr	1.265	1.140
	KBr	1.345	1.141
	RbBr	1.265	1.140
	CsBr	1.221	1.134
	RbCl	1.109	1.112
	CsCl	1.085	1.120
poly (His)	CsCl	1.210	1.350
	CsBr	1.442	1.382
	RbCl	1.238	1.359
	RbBr	1.471	1.366
	KBr	—[a]	1.360
	RbBr	1.471	1.366
	CsBr	1.427	1.382
	RbCl	1.238	1.359
	CsCl	1.210	1.350

[a] The ρ_0 of poly (His) at low pH exceeds the solubility of KBr.

to more fully understand the buoyant behavior of proteins in density gradients. The buoyant titrations discussed above provide considerable information as to the pH at which individual residues ionize and the associated $\Delta\rho$. However, in the case of the lysine and tyrosine residues, the titration occurs concomitantly and the separate contributions cannot be determined. It is of interest for all of the ionizable residues to determine the effect of replacing the ionizable moiety with another group of the same charge, of opposite charge, or with a neutral group. Two such studies have been made to date.

1. Carbamylation of Ovalbumin

The lysine residues of a protein can be converted to neutral homocitrulline residues by the addition of potassium cyanate at pH 9.0 (Stark and Smyth, 1963; Svendsen, 1967). This method was applied to ovalbumin (Ifft,

1971) and 18 of the 20 lysines converted to homocitrulline residues. The buoyant titration of this modified protein is contrasted with that of the native material in Figure 21. These data permitted the conclusions that follow.

The modification of the 18 residues contributed a density increment of 0.018 g/ml, or 0.001 g/ml for each residue. This is larger by a factor of 10 than the value predicted from the poly(Lys) data (0.040 g/ml × 20/377 = 0.002 g/ml). The titration curve of the modified material displays an inflection pH of 11.8, indicating that other species titrate in this region. The data were used to demonstrate that the $\Delta\rho$ at high pH is probably due to a combination of the loss of hydrated Cl^- and a gain of hydrated Cs^+.

2. Carbamylation of BMA

Studies parallel to those for ovalbumin have been completed for BMA (Ellis *et al.*, 1975). The titration curve for carbamylated BMA was 0.048 g/ml higher throughout the pH range 6–9 than the native curve. This difference decreased to 0.024 g/ml at high pH. These differences were caused by modification of only 25 out of the 58 lysine residues. As in the case of the ovalbumin data, homopolypeptide data were used to demonstrate that the $\Delta\rho_0$ observed between pH 10 and 12.7 was due to increased Cs^+ binding.

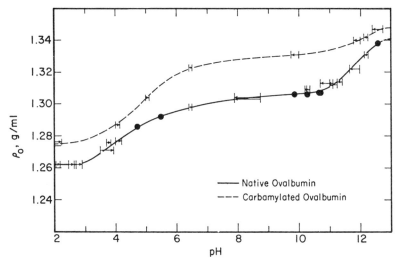

FIGURE 21. Buoyant densities of native and carbamylated ovalbumin as a function of pH; CsCl; 25°C. The arrows display the pH shifts with time, $pH_{initial} \rightarrow pH_{final}$. Runs involving no pH shifts are indicated by ● (Ifft, 1971).

E. Related Studies

In an effort to more fully understand the behavior of proteins in density gradients, several other research areas have been explored. The two studies discussed below have yielded the most information.

1. Potentiometric Titrations

The buoyant-titration plots shown above must be a reflection of the titration of the ionizable residues. It was of interest to determine directly the titration behavior of the proteins and polypeptides by measuring the deprotonation of these polymers as the pH is increased. These studies were required because potentiometric titrations in solutions of high CsCl concentrations are not available.

Potentiometric titrations in CsCl solutions have been measured for two proteins. The titration curve for ovalbumin was measured manually with a Radiometer pH meter (Ifft and Lum, 1971). The pH as a function of CsOH added was measured. Accurate determination of protein concentration by the Kjeldahl method permitted the standard plot of $\bar{h}$, the number of protons dissociated per molecule of isoelectric protein as a function of pH. The results of this titration are given in Figure 22. The observed titration behavior is

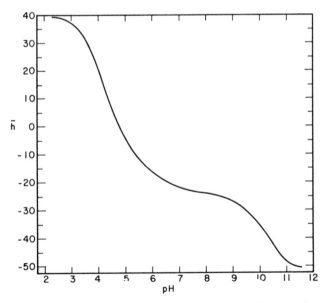

FIGURE 22. The number of protons associated per molecule of isoelectric ovalbumin $\bar{h}$, as a function of pH in 2.42 M CsCl at 25°C (Ifft and Lum, 1971).

quite standard in that a large inflection occurs at pH 4.3, where the carboxyls are expected to titrate, a modest inflection in the mid-pH range, where the histidines are deprotonated, and another large inflection near pH 10, where the lysines and tyrosines titrate.

The buoyant and potentiometric titration data could be combined by converting the data of Figure 22 to an α, the degree of ionization, vs. pH plot and plotting α vs. ρ_0. Figure 23 displays the results for ovalbumin. The plot is quite linear in the range $\alpha = 0.05$-0.65, which corresponds to the pH range 2.5–6.4. This is the region at which the carboxyl groups titrate. The linearity of the plot here indicates that the aspartic acid and glutamic acid residues all titrate with normal pK values and all make the same contribution to ρ_0. The pronounced bump in the curve at $\alpha = 0.65$ corresponds to the titration of the histidines. It suggests that the proposal of Rasper and Kauzmann (1962) that the histidines of ovalbumin titrate with an abnormally small volume change is correct.

Similar titrations have been completed for IgG (Ruark and Ifft, 1975). The Radiometer titration system was used to automatically record the reference and protein titration curves. The resultant α plot was combined with the buoyant titration data to yield α vs. ρ_0. This curve has about the same slope as the ovalbumin plot and was linear throughout most of the pH range.

Potentiometric titrations have been measured for all of the ionizable

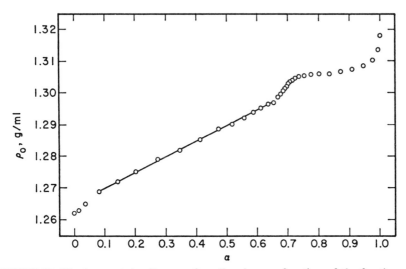

FIGURE 23. The buoyant density, ρ_0, of ovalbumin as a function of the fraction of residues, ionized, α, in CsCl solutions at 25°C (Ifft and Lum, 1971).

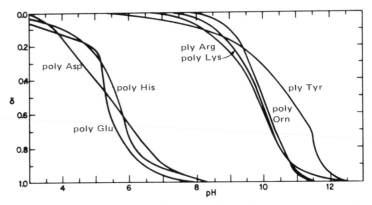

FIGURE 24. The fraction of residues deprotonated, α, of seven homopolypeptides as a function of pH in CsCl solutions of buoyant composition at 25°C (Almassy *et al.*, 1973).

homopolypeptides discussed earlier and several of the copolypeptides (Almassy *et al.*, 1973). Figure 24 shows the results for seven homopolypeptides. These data confirm the buoyant titration data in that there are two classes of residues, those which titrate at pH $\simeq$ 5.5 and those which titrate at pH $\simeq$ 10. It is to be noted that poly(Arg) does titrate with a pK of about 9.9, as was observed in the centrifuge studies. Few of these curves are smooth, ideal weak-acid titration curves due to the transition from one conformation to another and the accompanying phase change.

As in the case of ovalbumin and IgG, the potentiometric and buoyant data were compared. Table XII gives the midpoints of the inflections for each set of data. The midpoint-pH values compare rather well, with the

TABLE XII
pH at Midpoint of Transition

Homopolypeptide	Conc. CsCl[a] Buoyant	Conc. CsCl[a] Potentiometric	Low salt,[b] potentiometric
poly (Glu)	4.5	5.3	4.5
poly (His)	5.6	6.0	6.1
poly (Arg)	9.8	9.9	13.0
poly (Orn)	9.5	10.0	—
poly (Lys)	9.3	10.0	10.4
poly (Tyr)	10.5,12.2	10.6	9.5,11.5

[a] Almassy *et al.* (1973).
[b] Fasman (1967).

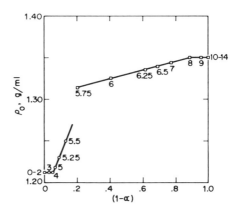

FIGURE 25. The buoyant density of poly(His) as a function of the degree of deprotonation in CsCl at 25°C: (○) soluble; (□) precipitate. Numbers next to the data points represent the pH.

exception of poly(Glu). Again, the anomalous behavior of poly(Arg) in concentrated CsCl solutions is apparent from the comparison with Fasman's value of pK_{int} (Fasman, 1967).

Figure 25 displays the ρ_0 vs. α plot for poly(His). It is observed that the low-pH, soluble form of the polypeptide behaves quite differently than the high-pH, uncharged precipitated form in that the slopes of the two lines differ by a factor of 10. However, in both cases there is a linear correlation between the degree of ionization and the buoyant density. The same behavior was observed for poly(Glu).

Sharp *et al.* (1975) have reported similar data for ionizable copolypeptides. Their α vs. ρ_0 data for poly(Glu$^{60.9}$Lys$^{39.1}$) also are linear when the degree of deprotonation is computed separately for the Glu and Lys residues.

2. Spectropolarimetry

We have recently acquired a Perkin Elmer Model 241 spectropolarimeter. It provides digital readout to 0.001° at four wavelengths. We have measured the rotation of BMA in water and in buoyant CsCl solutions as a function of pH (J. A. Geleris and Ifft, unpublished). The purpose of these studies is to determine whether significant structural changes occur in this protein in these concentrated salt solutions, especially at the extremes of pH. The data were converted to mean residue rotations in the standard manner. A value of 218 nm was selected for λ_0. The values of a_0 and b_0 obtained are displayed in Table XIII. Although some conformational changes are observed to occur above pH 9, the molecule retains most of its helical content up to the limits of the buoyant titration data, about pH 12. However, above this pH the molecule appears to lose most of its helical structure as the pH approaches 14. It is also apparent that the behavior of

TABLE XIII
Moffit Parameters, a_0 and b_0, for Native BMA
in Water and CsCl Solutions as a Function of
pH at 25°C

	In H_2O		In buoyant CsCl solutions	
pH	a_0	b_0	a_0	b_0
5.0	286	233	285	256
6.0	285	236	282	252
7.0	282	235	280	253
8.0	281	236	279	251
9.0	282	232	280	256
10.0	288	224	285	241
11.0	312	195	302	211
12.0	345	151	321	166

BMA as measured by the a_0, b_0 values is the same in water and buoyant CsCl solutions.

V. REFERENCES

Almassy, R., Zil, J. S. V., Lum, L. G., and Ifft, J. B. (1973) *Biopolymers* **12**: 2713.
Baldwin, R. L. (1959) *Proc. Natl. Acad. Sci. U.S.A.* **45**: 939.
Casassa, E. F. and Eisenberg, H. (1961) *J. Phys. Chem.* **65**: 427.
Cohen, G. and Eisenberg, H. (1968) *Biopolymers* **6**: 1077.
Cohn, E. J. and Edsall, J. T. (1943) *Proteins, Amino Acids and Peptides as Ions and Dipolar Ions*, Reinhold, New York.
Cox, D. J. and Schumaker, V. N. (1961) *J. Am. Chem. Soc.* **83**: 2439.
Eisenberg, H. (1967) *Biopolymers* **5**: 681.
Ellis, D. A., Coffman, V., and Ifft, J. B. (1975) *Biochemistry* **14**: 1205.
Fasman, G. D. (1967) *Poly-α-Amino Acids*, Marcel Dekker, New York.
Fujita, H. (1962) *Mathematical Theory of Sedimentation Analysis*, p. 258, Academic Press, New York.
Goldberg, R. J. (1953) *J. Phys. Chem.* **57**: 194.
Hade, E. P. K. and Tanford, C. (1967) *J. Am. Chem. Soc.* **89**: 5034.
Hearst, J. E. and Schmid, C. W. (1973) *Meth. Enzymol.* **XXVII** (Part D): 111.
Hearst, J. E. and Vinograd, J. (1961a) *Proc. Natl. Acad. Sci. U.S.A.* **47**: 999.
Hearst, J. E. and Vinograd, J. (1961b) *Proc. Natl. Acad. Sci. U.S.A.* **47**: 1005.
Hearst, J. E., Ifft, J. B., and Vinograd, (1961) *Proc. Natl. Acad. Sci. U.S.A.* **47**: 1015.
Hu, A. S. L., Bock, R. M., and Halvorson, H. O. (1962) *Anal. Biochem.* **4**: 489.
Hvidt, A., Johansen, G., Linderstrøm-Lang, K., and Vaslow, F. (1954) *C. R. Trav. Lab. Carlsberg* **29**: No. 9.

Ifft, J. B. (1969) in *A Laboratory Manual of Analytical Methods of Protein Chemistry* (P. Alexander and H. P. Lundgren, eds.) Vol. 5, p. 151, Pergamon Press, New York.

Ifft, J. B. (1971) *C. R. Trav. Lab. Carlsberg* **38**: 315.

Ifft, J. B. (1973) *Meth. Enzymol.* **XXVII** (Part D): 128.

Ifft, J. B. and Lum, L. G. (1971) *C. R. Trav. Lab. Carlsberg* **38**: 339.

Ifft, J. B. and Vinograd, J. (1962) *J. Phys. Chem.* **66**: 1990.

Ifft, J. B. and Vinograd, J. (1966) *J. Phys. Chem.* **70**: 2814.

Ifft, J. B. and Williams, A. (1967) *Biochim. Biophys. Acta* **136**: 151.

Ifft, J. B., Voet, D. H., and Vinograd, J. (1961) *J. Phys. Chem.* **65**: 1138.

Ifft, J. B., Martin, W. R., III, and Kinzie, K. (1970) *Biopolymers* **9**: 597.

Kuntz, I. D., Jr. and Kauzmann, W. (1974) *Adv. Protein Chem.* **28**: 239.

Linderstrøm-Lang, K. and Lanz, H., Jr. (1938) *C. R. Trav. Lab. Carlsberg* **21**: No. 24.

Linderstrøm-Lang, K., Jacobsen, O., and Johansen, G. (1938) *C. R. Trav. Lab. Carlsberg* **23**: No. 3.

Mandel, M. (1972) in *Handbook of Biochemistry, Selected Data for Molecular Biology.* 2nd ed. (H. A. Sober, ed.) pp. H-9–H-11. The Chemical Rubber Co., Cleveland.

Meselson, M. and Stahl, F. W. (1958) *Proc. Natl. Acad. Sci. U.S.A.* **44**: 671.

Meselson, M., Stahl, F. W., and Vinograd, J. (1957) *Proc. Natl. Acad. Sci. U.S.A.* **43**: 581.

Rasper, J. and Kauzmann, W. (1962) *J. Am. Chem. Soc.* **84**: 1771.

Ruark, J. E. and Ifft, J. B. (1975) *Biopolymers*, in press.

Scatchard, G., Coleman, J. S., and Shen, A. L. (1957) *J. Am. Chem. Soc.* **79**: 12.

Schildkraut, C. L., Marmur, J., and Doty, P. (1962) *Biol.* **4**: 430.

Schmid, C. W. and Hearst, J. E. (1969) *J. Mol. Biol.* **44**: 143.

Schmid, C. W. and Hearst, J. E. (1971) *Biopolymers* **10**: 1901.

Sharp, D. S., Almassy, R. J., Lum, L. G., Kinzie, K., Zil, J. S. V., and Ifft, J. B. (1975) In preparation.

Stark, G. R. and Smyth, D. G. (1963) *J. Biol. Chem.* **238**: 214.

Sueoka, N., Marmur, J., and Doty, P. (1959) *Nature* **183**: 1429.

Svendsen, I. (1967) *C. R. Trav. Lab. Carlsberg* **36**: 235.

Trautman, R. (1960) *Arch. Biochem. Biophys.* **87**: 289.

Vinograd, J. and Hearst, J. E. (1962) *Fortschr. Chem. Org. Naturst.* **XX**: 372.

Williams, A. E. and Ifft, J. B. (1969) *Biochim. Biophys. Acta* **181**: 311.

Williams, J. W., Van Holde, K. E., Baldwin, R. L., and Fujita, H. (1958) *Chem. Rev.* **58**: 715.

HOLLOW-FIBER   8
SEPARATION DEVICES
AND PROCESSES

BURTON A. ZABIN

I. INTRODUCTION

As proteins and other biologically active molecules are highly sensitive to
phase changes, the process used to separate and concentrate them should
not require any harsh treatment that might deactivate the molecule. One
such mild method is the hollow-fiber process, a molecular-sieving membrane
technique that separates by differences in molecular size and configuration.

The hollow-fiber process is fundamentally similar to other molecular
sieving methods—porous gels (dextran or polyacrylamide spheres containing
pores of controlled dimensions) and flat membranes with similar properties.
In all three methods the process of separation is essentially the same: molecules larger than the pores of the molecular sieve will be excluded, while
molecules smaller than the pores will pass through, and the rate of passage
will be determined by their molecular size and configurations. This means
that a mixture of molecules, all smaller than the pores of the sieve, can still
be separated because they will pass through the pores at different rates. This
separation process is shown graphically in Figure 1.

BURTON A. ZABIN, Bio-Rad Laboratories, Richmond, California, 94804.

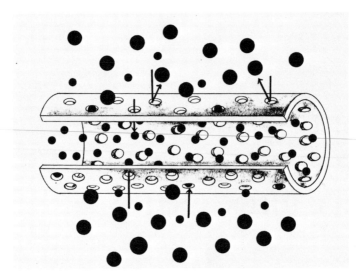

FIGURE 1. Artist's concept of a section of a hollow-fiber membrane. Molecules larger in size than the pores of the fiber are excluded, while molecules smaller in size than the pores permeate through the membrane.

In Figure 2, an actual hollow-fiber membrane is shown under 750 × magnification. These membranes are about 200 μm in diameter with about 25-μm walls. When a large number of these fibers are sealed into a convenient device so that the fibers can be filled with sample or solute, as shown in Figure 3, they provide a very high surface area in a very small, compact space. In fact, a hollow-fiber device measuring only 4 inches in height contains the same surface area as 80 flat membranes of the common 4-cm diameter. This makes the hollow-fiber devices extremely efficient, generally more convenient than columns, and less space-requiring than flat membranes.

In dialysis, a solution of varying-molecular-weight solutes is placed inside the fibers and an aqueous solution is placed outside and surrounding the fibers. Solutes, such as salt, smaller than the pore size of the fibers will diffuse through the fiber walls, leaving the larger molecules behind inside the fibers (Figure 4). The rate of permeation through the fiber walls is inversely proportional to molecular size, the smaller molecules permeating at a greater rate than the larger molecules. This makes it possible to fractionate solutions containing several solutes of varying molecular weight.

When a pressure gradient is established, not only will the low-molecular-weight solutes pass through the walls of the fibers, but the solvent will pass through as well. In this way, the larger molecules in the solution will be concentrated.

<table>
<tr><td>25 MICRONS</td><td>750 X</td></tr>
</table>

FIGURE 2. Photomicrograph (750 ×) of a section of hollow fiber.

II. HOLLOW-FIBER MEMBRANES

A variety of hollow-fiber membranes are available today. One type called "Bio-Fibers" (trademark of The Dow Chemical Company) are derived from cellulose and cellulose acetate. These fibers are homogeneous or isotropic in structure, permitting solvent flow in either direction, i.e., from inside the fibers to the outside container or in the reverse direction. Molecular-weight cutoff is the major property that distinguishes each of the Bio-Fiber types from another. Nominal molecular-weight cutoff values for each

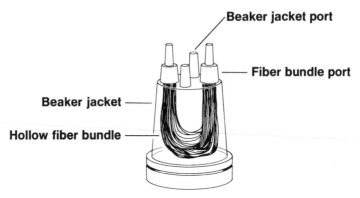

FIGURE 3. Schematic diagram of a hollow-fiber device.

fiber type have been established by determining the molecular weight of that solute which is 85% rejected by the membrane on the basis of the following relationship:

$$\% \text{ Rejection} = 100 \times \frac{\text{(Solute) Retentate} - \text{(Solute) Filtrate}}{\text{(Solute) Retentate}}$$

Figure 5 illustrates the effect of solute molecular weight on percent rejection using Bio-Fiber membranes.

Table I lists the chemical composition of Bio-Fiber membranes and the nominal molecular-weight cutoffs derived from the above curve. Bio-Fiber 5,

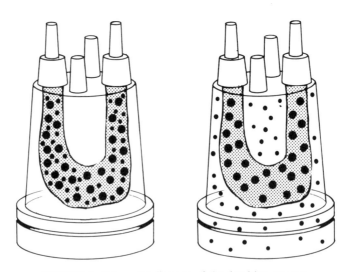

FIGURE 4. Schematic diagram of the desalting process.

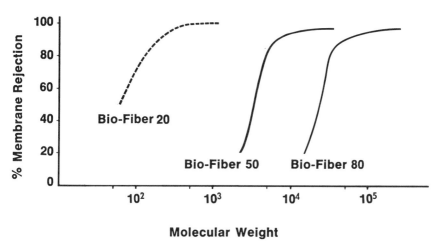

FIGURE 5. Effect of solute molecular weight on percent rejection of Bio-Fiber 80 and Bio-Fiber 50. The Bio-Fiber 20 curve is for nonionic solutes according to molecular weight.

a silicone rubber membrane used for the separation of gases by differential rates of diffusion, has no listed cutoff value.

III. DEVICE CONFIGURATIONS

Bio-Fiber devices are designed to accommodate a wide range of sample volumes with equal ease of operation. Figure 6 shows five device configurations for processing sample volumes from a few milliliters to several liters.

TABLE I
Properties and Applications of Bio-Fiber Membranes

Fiber type	Chemical composition	Nominal molecular-weight cutoff	Applications[a] First	Second
Bio-Fiber 80	Cellulose acetate	30,000	C	D
Bio-Fiber 50	Cellulose	5,000	D	C
Bio-Fiber 20	Cellulose acetate	200	C	D [b]
Bio-Fiber 5	Silicone rubber	—	Gas transfer	

[a] C = concentrating, D = dialyzing.
[b] Nonionic solutes only.

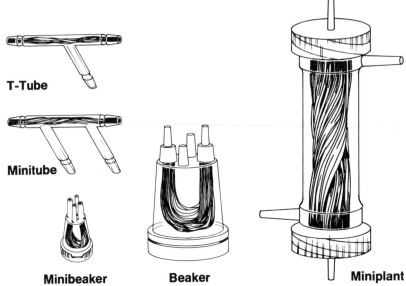

T-Tube

Minitube

Minibeaker   **Beaker**   **Miniplant**

FIGURE 6. Hollow-fiber-device configurations.

the hollow fibers are permanently mounted inside the devices in compact fiber bundles.

The devices are designed so that solutions can flow both inside the fibers through fiber ports and outside the fibers through device ports.

The Bio-Fiber Minitube is a straight plastic tube with two side arms. A bundle of hollow fibers is sealed in the straight portion. The two side arms are used to pass sample solution or dialysis solution into and out of the area surrounding the fiber bundle (jacket area).

The Bio-Fiber Minitube can be used for dialyzing small sample volumes by filling the Minitube jacket with sample solution and passing a dialysis solution through the fiber bundle. The small solutes in the sample will diffuse into the fibers and be carried away in the dialysis solution. It is also possible to dialyze a sample solution by reversing the placement, putting the sample inside the fibers with the dialysis solution in the Minitube jacket.

Sample volumes larger than the Minitube fiber-bundle volume or the jacket volume can be dialyzed either by gravity flow from a reservoir or by recycling with a pump.

The Minitube can be used to concentrate samples by filling either the jacket or the fibers with sample and creating a pressure differential across the fiber membrane. Bio-Fiber 80, 50, 20, and 5 are available in the Minitube configuration.

The Bio-Fiber T-Tube is a *T*-shaped tube with a straight bundle of fibers sealed inside. Because of its small size, configuration, and low fiber-

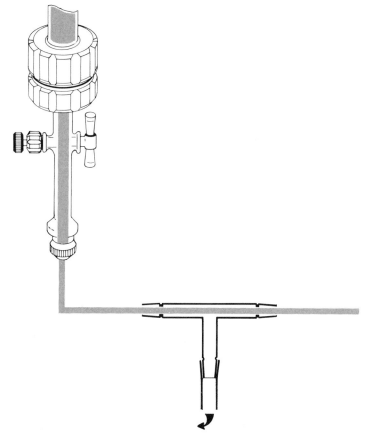

FIGURE 7. Concentrating a column effluent using a Bio-Fiber T-Tube. The column effluent passes through the fiber bundle. Vacuum is applied to the T-Tube jacket.

bundle volume, the T-Tube is used almost exclusively for concentrating column effluents by passing the sample through the fiber-bundle—in much the same way a sample is passed through a chromatography column—while applying a vacuum to the side arm (jacket port). Bio-Fiber 80 is the only fiber type available in the T-Tube configuration. Figure 7 illustrates a typical setup for concentrating a column effluent.

The Minibeaker is similar to the Minitube in fiber-bundle volume but larger in jacket volume. The Minibeaker is different from the Minitube in that the fiber bundle is U-shaped and is sealed into a small, upright, beaker configuration. The Minibeaker is fitted with a removable bottom which provides an area for a magnetic stir flea. The two fiber ports and the two jacket ports are located on top of the device.

The Minibeaker is used primarily for dialyzing biological solutions in a manner similar to the Minitube. The Minibeaker is available only with Bio-Fiber 50 fibers.

The Bio-Fiber Beaker is used for dialyzing and concentrating sample solutions of moderate volume. The Beaker is similar to the Minibeaker but has 10 times the fiber-bundle capacity and 12 times the jacket capacity. It is also fitted with a removable bottom which can accommodate a magnetic stir bar. A plastic mesh screen separates the fiber-bundle area from the stir-bar area to prevent possible stir-bar damage to the fibers.

The Beaker can be used to dialyze, concentrate, or concurrently dialyze and concentrate sample solutions. For dialysis, if the sample is placed inside the fibers, then the dialysis solution is placed in the Beaker jacket. If the sample is placed in the Beaker jacket, then the dialysis solution is placed in the fibers.

The Beaker can be used to concentrate samples by filling the Beaker jacket with sample and then creating a pressure differential across the fiber membrane. This is generally done by applying a vacuum to one of the fiber outlets with the other fiber outlet capped. Low-molecular-weight solutes and solvent pass through the fiber walls into the fibers themselves and are drawn by the vacuum into a vacuum trap. The Beaker can dialyze or concentrate sample volumes larger than the Beaker jacket volume by recycling with a pump or by gravity flow from a reservoir.

The Bio-Fiber Beaker is available with either Bio-Fiber 80, 50, 20, or 5 fibers.

The Bio-Fiber Miniplant is the largest Bio-Fiber device now available. The Miniplant is used for dialyzing or concentrating large sample volumes. It is used in the same way as the Minitube and Beaker. The Miniplant is available with either Bio-Fiber 80, 50, or 20 fibers.

IV. DIALYSIS

Bio-Fiber devices can be used for:

1. Removal of inorganic salts, i.e., desalting.
2. Removal of low-molecular-weight impurities, i.e., purifying.
3. Exchange of low-molecular-weight salts, i.e., buffer exchange.

All these operations are done in essentially the same way. Salts and other low-molecular-weight components diffuse from the sample on one side of the fibers to the dialysis solution on the other side of the fibers. Because the fibers are isotropic, solutes can diffuse through the fibers in either direc-

TABLE II
Bio-Fiber Device Capacities

Device	Fiber-bundle volume, ml	Jacket volume, ml
Minitube	0.3	2.5
T-tube	0.3	2.0
Minibeaker	0.4	8
Beaker	4	100
Miniplant	150[a]	100
	120[b]	150

[a] Bio-Fiber 80, 20.
[b] Bio-Fiber 50.

tion, and, from that standpoint, it makes no difference if the sample is placed inside the jacket or inside the fibers. That decision is based on the nature and volume of the sample and the configuration of the device that is being used.

For most dialysis applications, fiber selection is dictated by the molecular weight of the solute of interest. The device selection is based on the sample volume. Fiber selection for dialysis is easy because for most dialysis applications Bio-Fiber 50 is the best choice. There are two general exceptions: Bio-Fiber 20 is the best choice for dialyzing low-molecular-weight, nonionic molecules such as sucrose, urea, or 2-mercaptoethanol without loss of ionic molecules, and Bio-Fiber 80 provides more rapid dialysis of larger-molecule-weight impurities (approximately 200 mol. wt. or more) than does Bio-Fiber 50 because of the former's larger pore size.

TABLE III
Recommended Sample Flow Rates for Single-Pass
Dialysis with Bio-Fiber 50

Device	Sample flow rate[a] for NaCl removal, ml/min		
	99%	90%	75%
Minitube	0.05	0.01	0.2
Minibeaker	0.05	0.15	0.3
Beaker	0.4	0.8	1.5
Miniplant[b]	90	200	400

[a] Sample flow through fiber bundle, dialysis-solution flow through device jacket at 5–20 times sample flow rate.
[b] Figures for Miniplant at 37°C, balance of devices at 25°C.

Selection of the device is based almost entirely on sample size. Each device has two capacities: the fiber-bundle volume (capacity inside the fibers) and the jacket volume (capacity outside the fibers). This means there is an overlap in capacity from one device to another, as shown in Table II. As a general rule, it is better to place the sample outside the fibers. A solution can be dialyzed by a single pass through a Bio-Fiber device. In general, the slower the sample flow rate the higher the dialysis efficiency. Table III lists the recommended sample flow rates for Bio-Fiber 50 single-pass procedures.

V. COMBINED DIALYSIS AND CONCENTRATION

The properties of Bio-Fiber membranes permit the permeation of low-molecular-weight solutes (dialysis) or solvent (concentration) or, under proper conditions, the permeation of both solute and solvent. The simultaneous dialysis and concentration of biological compounds may be desirable or required for specific separations. As previously discussed, solutions may be dialyzed by placing the sample on one side of the membrane while passing dialysis solution on the opposite side of the membrane. Most Bio-Fiber devices can concentrate at the same time when a pressure differential is generated across the fiber membrane. Operationally, solutions can be desalted and concentrated by drawing water from the device at a rate slightly greater than the rate at which water is being added. For example, if a sample is placed in a Beaker jacket and water is drawn from the fiber-bundle outlet at 30 ml/min while adding water to the fiber bundle inlet at 29 ml/min, the sample in the Beaker jacket will be concentrated at a rate of 1 ml/min.

Techniques and procedures for separating, isolating, and purifying a biological solute often call for an exchange of buffer, usually a time-consuming and complicated procedure when traditional techniques are used. The isotropic property of Bio-Fiber membranes combined with their in-

TABLE IV
Relative Solution Retention for Bio-Fiber Membranes[a]

Solute mol. wt.	Bio-Fiber 80	Bio-Fiber 50	Bio-Fiber 20
100,000	High	High	High
60,000	Moderate	High	High
30,000	Low	High	High
10,000	Not recommended	Moderate	High
5,000	Not recommended	Low	High
200	Not recommended	Not recommended	Moderate

[a] High (90–100%), moderate (65–90%), low can be used, but solute permeation will reduce recovery.

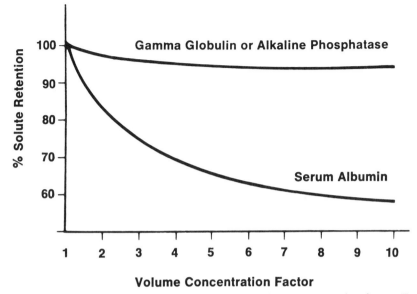

FIGURE 8. Percent of solute retained by fiber vs. volume-concentration factor. Top curve, gamma globulin (Bio-Fiber 80) and alkaline phosphatase (Bio-Fiber 50). Bottom curve, albumin (Bio-Fiber 80). Beaker device used for all determinations.

herent speed of operation make Bio-Fiber devices the method of choice for this operation.

VI. CONCENTRATION

When a pressure differential is established across the wall of a hollow-fiber membrane, low-molecular-weight solutes and solvent will pass through the fiber wall, leaving the larger solutes behind in a higher concentration. The differential can be established with either vacuum or pressure, depending upon the sample size and the device that is used. In concentrating with the Bio-Fiber devices, there are three parameters to deal with: the choice of fiber, the selection of the device, and the use of either pressure or vacuum.

For concentrating high-molecular-weight solutes Bio-Fiber 80 is recommended because of its high water flux rate and nominal molecular-weight cutoff of 30,000. Bio-Fiber 50 is recommended for concentrating intermediate-molecular-weight solutes with maximum retention because of its nominal molecular-weight cutoff of 5000. Bio-Fiber 20 can be used for concentrating very low-molecular-weight solutes by osmosis, using either 15% calcium chloride or 20% polyethylene glycol solution.

The choice between Bio-Fiber 50 and Bio-Fiber 80 depends upon the molecular weight of the solute being concentrated and the desired solute

TABLE V
Bio-Fiber Devices, Sample Volumes, and Water Flux Rates[a]

Bio-Fiber device	Sample volume, ml		Water flux, ml/min	
	Initial	Final	Bio-Fiber 80	Bio-Fiber 50[b]
Minitube	20	0.5	0.5	0.05
T-tube	20	0.5	0.5	NA[c]
Minibeaker	20	0.5	NA	0.05
Beaker	1000	5	7	0.7
Miniplant	1000	100–150	70	10

[a] Actual concentration rates will be less than water flux rates depending on solute concentration, composition, and temperature.
[b] Bio-Fiber 50 rates can be increased 2–3 times using 20% polyethylene glycol to remove water from the sample by osmosis.
[c] NA = not available.

recovery. Table IV shows the relative solute retention of these two fibers.

In addition to solute size, the volume-concentration factor (ratio of initial concentration to final concentration) also affects solute retention.

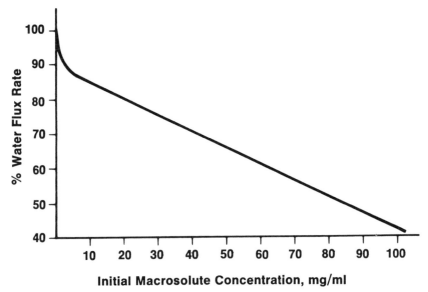

FIGURE 9. Effect of dissolved macrosolute on water flux rate. Obtained using bovine serum albumin in fiber bundle of Bio-Fiber 80 Beaker.

Figure 8 illustrates the retention of gamma globulin, serum albumin, and alkaline phosphatase.

The choice of the device depends almost entirely on sample size. Table V shows the maximum initial sample volume and the minimal final concentrated volume, plus the water flux rates, for all of the Bio-Fiber devices. The composition and concentration of solutes also affect concentration rates. Figure 9 shows the effect of dissolved protein on Bio-Fiber 80 flux rates. For example, an initial concentration of 30 mg/ml will be concentrated at 75% of the water flux rate, or about 45 ml/min for a Miniplant.

In some instances researchers have run into problems in obtaining high recoveries using membrane techniques because biological molecules tend to "stick" to membrane surfaces. Two different types of adsorption can be observed: chemical adsorption and physical adsorption.

Chemical interaction can be observed simply by contacting a solution with membrane under static condition. Table VI shows the adsorption of some biologicals onto Bio-Fiber membranes. The relatively inert nature of cellulose and cellulose acetate used in the Bio-Fiber membranes results in minimum chemical adsorption.

When a pressure differential is established across a membrane during a concentration process, a physical adsorption occurs. This adsorption, or "solute polarization," results in a protein layer being deposited on the membrane surface. Considering the high surface area of the Bio-Fiber devices and the dilute nature of many biological solutions, this solute polarization effect can be pronounced, with a significant percentage of the molecules "stuck" to the membrane surfaces. Fortunately, this phenomenon can be overcome. Bio-Fiber membranes are isotropic, so the solute can flow through them in either direction. This fact can be applied to a simple backwashing technique that will recover the "stuck" molecules and yield a high recovery.

TABLE VI
Adsorption of Biologicals onto Various Membranes[a]

Membrane type	Adsorption, g/cm²		
	Vitamin B-12	Cytochrome c	Albumin
Bio-Fiber 50[b]	0.28	0.1	1.0
Bio-Fiber 80[c]	0.61	0.61	1.4

[a] The solutions, in 1000-mg/liter concentrations, were buffered at pH 7.0 and exposed to the membranes for 24 hr at 25°C.
[b] Cellulose composition.
[c] Cellulose acetate.

To backwash, simply reverse the direction of solvent flow and run two times the fiber-bundle volume through the fibers. The backwash technique is most effective when the sample has been concentrated in the jacket of the device. Backwashing will increase sample recoveries as much as 50%.

VII. DRUG-BINDING STUDIES

The Minibeaker Dialyzer can be used as an equilibrium cell for the determination of free and protein-bound species. Small samples can be loaded into the fibers by attaching 1.5-inch length tubing to one fiber port, adding sample into the tubing, and centrifuging at 2000 g for 5 min.

VIII. ENZYME REACTOR

Enzymes are successfully contained in (and, in effect, "insolubilized" within) Bio-Fibers with no detectable enzyme leakage. Alkaline phosphatase (0.1 mg/ml) in 0.05 M tris, pH 8.0, was placed inside the fibers of a Bio-Fiber 50 Beaker. Reaction with disodium 4-nitrophenol phosphate (100 ml) placed in the jacket showed that the initial reaction rate is directly proportional to the substrate concentration (36–640 μM) (see Appendix). This results in the following advantages:

1. Separation of enzyme and product will occur *in situ* as the reaction proceeds.
2. Substrates which cannot permeate the membrane cannot react with the enzyme to form products.
3. Products acting as negative feedback agents may be removed as they are found.
4. The enzyme solution within the fiber bundle may be recycled, purified, or regenerated outside the reactor.

IX. CONCLUSION

In conclusion, the development of hollow fibers and the hollow-fiber devices has brought a new level of efficiency and ease of operation to the separation of proteins and other biologically active molecules.

X. APPENDIX

Typical Bio-Fiber Applications

Sample	Bio-Fiber application	Fiber type	Reference*
Albumin	Concentration	80	27
Albumin	Dialysis	80	27
Alkaline phosphatase	Concentration	50	4
Alkaline phosphatase	Enzyme reactor	50	9, 22
Ammonium sulfate removal	Dialysis	50, 80	6
ATP-dependent DNase	In-line dialysis	50	21
ATP-dependent DNase	Concentration	80	21
Unreacted ATP removal	Buffer exchange	50	20
Bacillus subtilis filtrate	Concentration	80	18
Cancer cell culture	Gas permeation	5	17
Artificial capillary	Microsurgery	80	19
Catalase	Buffer exchange	50	23
—Chymotrypsin	Equilibrium dialysis	50	23
Colleototriphum–Lindemuthian filtrate	Concentration	80	1
Colloidal particles	Concentration	80	6
Creatinine phosphokinase isoenzyme	Concentration	80	24
Creatinine phosphokinase isoenzyme	Dialysis	80	24
Tritiated dimethylbenzylrifampicin	Dialysis	50	16
Calf thymus DNA	Concentration	80	4
Mouse embryo palate	Organ perfusion	5	7
Enzyme extract	Dialysis/concentration	80	5
Equine gamma golubin	Concentration	50	6
20-Hydroxysteroid dehydrogenase	Dialysis	50	25
IgG	Concentration	50	2
IgG	Dialysis	50	2
Goat IgG	In-line dialysis	50	15
Potato invertase inhibitor	Protein-free filtration	80	26
4 M 2-Mercaptoethanol removal	Dialysis	20	6
Periplasmic proteins	In-line concentration	50	11
Horseradish peroxidase isoenzyme	Concentration	50	12
Protein (50,000 mol. wt.)	Concentration	50	13
Prothrombin complex	Concentration	50	14
Human serum	Dialysis	50	6
Chicken triose phosphate isomerase	Concentration	50	10
Tritium–hydrogen exchange	Dialysis	50	8
Triton X-100 removal	Dialysis	50	16
8 M Urea removal	Dialysis	20	6
Virus (polio type I)	Concentration	80	3

*1. Anderson, A. (1974) University of Colorado, Boulder, Colo. 80220, Personal communication.
 2. Angellety, L. (1974) Electro Nucleonics, Bethesda, Md. 20014, Personal communication.
 3. Belfort, G. (1974) Human Environmental Sciences Program, Hebrew University, Jerusalem, Personal communication.

4. Bio-Rad Laboratories (1973) Bulletin 2004, *Concentrating Biological Solutions.*
5. Bio-Rad Laboratories (1973) Bulletin 2005, *Enzyme Dialysis and Concentration.*
6. Bio-Rad Laboratories (1973–1974) Research results.
7. Brinkley, L. (1974) Department of Oral Biology, University of Michigan Dental School, Ann Arbor, Mich. 48104, Personal communication.
8. Browne, C. A. and Waley, S. G. (1973) *Anal. Biochem.* **56**: 289.
9. Davis, J. C. (1974) Kinetics Studies in a Continuous Steady-State Hollow Fiber Membrane Enzyme Reactor, *Biotech. and Bioeng.* **8**: 1113 (To be published).
10. Fischer, E. F. (1973) Department of Chemistry, University of California, La Jolla, Calif. 92037, Personal communication.
11. Furlong, C. E., Willis, R. C., and Morris, R. G. (1973) Department of Biochemistry, University of California, Riverside, Calif. 92502, Personal communication.
12. Girotti, A. W. (1974) Biochemistry Department, Medical College of Wisconsin, Milwaukee, Wisc. 53202, Personal communication.
13. Granger, G. A. (1973) Department of Molecular Biology, University of California, Irvine, Calif. 92664, Personal communication.
14. Heavilon, L. (1974) American Red Cross, Lansing, Mich. 48924, Personal communication.
15. Holladay, D. W. (1973) Oak Ridge National Laboratory, Oak Ridge, Tenn. 37830, Personal communication.
16. Joss, U. and Calvin, M. (1973) Laboratory of Chemical Biodynamics, University of California, Berkeley, Calif. 94720, Personal communication.
17. Knazek, R. A., Gullino, P. M., Kohler, P. O., and Dedrick, R. L. (1972) *Science* **178**: 65.
18. Lababitch, J. (1974) Chemistry Department, University of Colorado, Boulder, Colo. 80220, Personal communication.
19. Matioli, G. (1974) Microbiology Department, University of Southern California, School of Medicine, Los Angeles, Calif. 90033, Personal communication.
20. Mayer, S. E. (1973) Division of Pharmacology, University of California, San Diego, La Jolla, Calif. 92037, Personal communication.
21. Ohi, S. and Sueoka, N. (1973) *J. Biol. Chem.* **248**: 7336.
22. Rony, P. R. (1972) *J. Am. Chem. Soc.* **94**: 8247.
23. Smith, R. N. and Hansch, C. (1973) *Biochemistry* **12**: 4924.
24. Sobel, B. E. (1974) Cardiology Department, Washington University School of Medicine, St. Louis, Mo. 63110, Personal communication.
25. Sweet, F. (1973) Department of Obstetrics, Washington University School of Medicine, St. Louis, Mo. 63110, Personal communication.
26. Varns, J. (1973) Red River Valley Potato Research Laboratory, East Grand Forks, Minn. 56721, Personal communication.
27. Weiss, E. (1974) Cardiology Department, Washington University School of Medicine, St. Louis, Mo. 63110, Personal communication.

AFFINITY CHROMATOGRAPHY, PRINCIPLES AND APPLICATIONS

9

INDU PARIKH AND PEDRO CUATRECASAS

Since the time the term "affinity chromatography" was first coined a few years ago (Cuatrecasas *et al.*, 1968), the basic idea has been widely exploited as a powerful tool for the separation and purification of a wide variety of biological macromolecules. Its effectiveness for purification rests on the selectivity of interaction, and thus of adsorption, of a biological macromolecule on an affinity adsorbent which is prepared by the covalent immobilization of a specific ligand on a solid polymeric matrix. In the case of an enzyme, an appropriate reversible competitive inhibitor is immobilized. Substrates or cofactors of the enzyme may also be used under selected experimental conditions (Cuatrecasas, 1970; Cuatrecasas and Anfinsen, 1971). The desorption (i.e., elution) of the macromolecule from the affinity column is achieved either by perturbing the interaction between the macromolecule and the adsorbent, or by including a competing ligand in the eluting buffer. For obvious reasons, the design of a new affinity-chromatographic system for a given macromolecule requires individual attention in the selection and

INDU PARIKH and PEDRO CUATRECASAS, Departments of Pharmacology and Experimental Therapeutics and Medicine, The Johns Hopkins University School of Medicine, Baltimore, Maryland 21205.

attachment of the ligand as well as selection of the buffer conditions for the adsorption and desorption processes. In principle, virtually any specifically interacting system composed of two or more species can be approached by this method.

In most instances of affinity chromatography the physical parameter of greatest importance is the dissociation constant. The magnitude of this quantity determines whether the system will have a strong enough association to be useful, and it can dictate how the adsorbed biological macromolecule will need to be treated to recover it from the affinity column. Many of the basic principles behind affinity chromatography have been derived from experience with the purification of staphylococcal nuclease. p-Aminophenyl-3′,5′-thymidine diphosphate is a competitive, reversible inhibitor of staphylococcal nuclease with a K_m of 10^{-6} M. Affinity columns made by covalently attaching the inhibitor to agarose are able to selectively adsorb and retain this enzyme from crude preparations if the pH, ionic strength, and metal requirements for binding are fulfilled. The contaminating proteins are removed unretarded. The pure enzyme can then be eluted by changing the buffer conditions so that the strength of the interaction (i.e., affinity) between the enzyme and the immobilized inhibitor is diminished or destroyed. In the case of staphylococcal nuclease, 0.1 N acetic acid is used to elute the pure, homogeneous enzyme in a single sharp peak (Figure 1). Other examples of enzyme purification by affinity chromatography have been included under appropriate sections.

Peptides. Besides purification of enzymes, which continues to be the primary application, affinity chromatography has been successfully used in the isolation of synthetic peptides such as some synthetic fragments of staphylococcal nuclease (Ontjes and Anfinsen, 1969; Parikh *et al.*, 1971) as well as ribonuclease (Kato and Anfinsen, 1969). This has been possible because of the interesting observation that certain polypeptide fragments of these enzymes exhibit noncovalent interaction in properly buffered solutions with one of the other polypeptide fragments of the same enzyme. If such peptide–peptide complementation results in a sufficiently stable complex, affinity chromatography can obviously be exploited to isolate one or the other peptide fragment from a crude synthetic mixture. The technique has permitted isolation of synthetic peptides in high yields. Affinity chromatography may also be used for the isolation of affinity-labeled peptides (Figure 2), which are obtained after active site labeling of an enzyme followed by tryptic or chymotryptic digestion.

If the affinity-labeled peptide has some residual affinity (because of the attached ligand) for the native enzyme, it can be isolated with an affinity column prepared by immobilizing the enzyme to agarose. Alternatively, antibodies against the active site-directed affinity-labeling reagent can be used (as an antibody–agarose column) to isolate the labeled peptide. The method

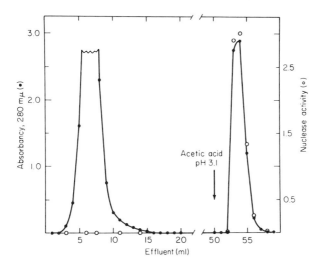

FIGURE 1. Purification of staphylococcal nuclease by affinity chromatography on a *p*-aminophenyl thymidinebiphosphate–agarose column (0.8 × 5.0 cm). The column was equilibrated with 0.05 M borate buffer, pH 8.0 containing 0.01 M CaCl$_2$. Approximately 40 mg of partially purified proteins containing 8 mg of nuclease in 3 ml of the above buffer was applied to the column. After washing the column with 40 ml of the buffer, the pure enzyme was eluted with 0.1 M acetic acid in almost quantitative yield. Data from Cuatrecasas *et al.* (1968).

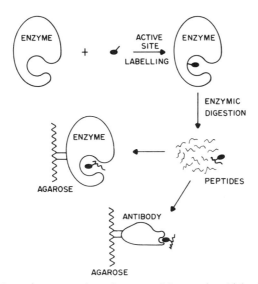

FIGURE 2. A schematic presentation of two possible ways in which affinity chromatography could be used in the isolation of specifically labeled peptides.

was ingeniously illustrated during isolation of DNPS-labeled (Wilchek and Miron, 1972) and nitrotyrosine peptides (Helman and Givol, 1971) of serum albumin and lysozyme, respectively.

Protein–Protein Complementarity. A not so well exploited application of affinity chromatography is illustrated by proteins which exhibit noncovalent interaction with other proteins. This type of protein–protein interaction may or may not have physiological significance. Table I illustrates a few examples of such types of interactions and their application in affinity chromatography.

Sulfhydryl Proteins. Organomercurial agarose has been used successfully in the purification of sulfhydryl-group-containing proteins. *p*-Chloromercurobenzoic acid is coupled to aminoagarose with water-soluble carbodiimide (Cuatrecasas, 1970). The organomercurial agarose has a very high capacity for SH-group-containing proteins. Papain has been elegantly purified on organomercurial agarose prepared by coupling *p*-aminophenylmercuric acetate by the cyanogen bromide activation method (Sluytermann and Wijdenes, 1970). There are various biological important SH-group-containing enzymes, such as creatine phosphokinase and phosphofructokinase, which could be purified through the use of such organomercurial gels. The adsorbed protein can be eluted easily from the affinity column with buffers containing chelating agents or various thiol reagents. The organomercurial agarose is readily regenerated by washing the column with a solution of $HgCl_2$.

TABLE I
Use of Protein–Protein Complementarity in the Affinity Chromatography

Agarose linked protein	Complementary protein purified	Reference
Neurophysin	Lys-vasopressin	Fressinaud *et al.*, 1973; Pradelles *et al.*, 1972
α-Lactalbumin	Lactose synthetase	Andrews, 1970; Trayer and Hill, 1971; Barber *et al.*, 1972
F-Actin	Tropomyosin and troponin	Konda *et al.*, 1972
Hapatoglobin	Hemoglobin	Javid and Liang, 1973; Klein and Mihaesco, 1973
Avidin	Acetyl-CO-A-carboxylase	Landman and Dakshinamurti, 1973
Trypsin	Soybean trypsin inhibitor and Limabean trypsin inhibitor	Robinson *et al.*, 1971; Feinstein, 1970; Chauvet and Acher, 1973; Foster and Ryan, 1974; Ohlsson and Tegner, 1973
Serum albumin	Ferriporphyrine peptide of cytochrome c	Wilchek, 1972

Mutant Enzyme Forms. β-Galactosidase, a multisubunit enzyme, has been isolated by affinity chromatography from microbial sources by Steers and coworkers (Steers *et al.*, 1971). These studies illustrate other aspects of affinity chromatography besides its isolation and purification. Steers *et al.* found that certain mutants of *E. coli* produce a catalytically inactive monomeric form of the enzyme instead of the normal tetrameric species. The monomeric form of β-galactosidase, though being devoid of any enzymic activity, is able to bind the specific enzyme inhibitor, *p*-aminophenyl-β-thiogalactoside, as demonstrated by the ability of the affinity column to purify such monomers. The ability of such monomers to hybridize with subunits of the wild-type enzyme, and thus to form a catalytically active native enzyme, further proves that the original defect is in subunit interaction rather than in the active site. Affinity chromatography might be useful in studies of various genetically based diseases where it is not known whether the enzyme involved is catalytically defective or altogether absent. Another aspect of the β-galactosidase studies shows how the protomer properties of an oligomeric enzyme can be studied in the absence of any denaturing agents. Figure 3 illustrates schematically how an enzyme monomer–agarose derivative can be prepared. Kinetic studies of subunit interactions with metal ions, substrates, or inhibitors, and also subunit–subunit interactions, are very difficult to perform in solutions because the subunit cannot be maintained in the dissociated form in the absence of denaturants. Special studies of mechanistic significance have also been performed with some other oligomeric proteins (Pass *et al.*, 1973; Lornitzo *et al.*, 1974; Feldmann *et al.*, 1972; Schmit *et al.*, 1970, Green and Toms, 1973; Green, 1973).

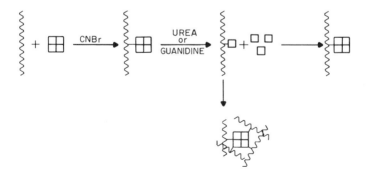

FIGURE 3. Protomer properties in oligomeric enzymes. Matrix-bound protomer unit of an oligomeric enzyme is prepared by dissociation of the immobilized intact enzyme with denaturants. Matrix-bound protomers are stable in contrast to free protomers, which tend to reassociate in the absence of denaturants. Reassociation of matrix-bound protomers has been suggested in one enzyme system.

Isolation of Viruses. There are relatively few reports on the application of affinity chromatography to the isolation of viral bodies. Influenza virus has been shown to adsorb specifically to an oxamic acid–agarose column, an affinity adsorbent for the enzyme neuraminidase (Cuatrecasas and Illiano, 1971). These results also show that this is a superficial viral enzyme. Influenza and Aleutian mink disease viruses are perhaps the best studied examples where affinity chromatography has been exploited successfully (Kenyon *et al.*, 1973). These viruses are isolated on affinity adsorbents to which a viral antibody has been covalently attached. The major limitation in the application of affinity chromatography for viruses is presumably the difficulty encountered in raising sufficient quantities of specific viral antibodies.

Estrogen Receptors. Affinity chromatography is proving rewarding in the study and isolation of complex cellular structures where conventional purification procedures are most difficult to apply. For example, the difficulties encountered in the isolation of estrogen receptors by conventional methods (Sica *et al.*, 1973) include the lack of availability of sufficient quantities of the receptor from any given tissue and the labile character of the receptor to relative extremes of pH and ionic strength. Perhaps the most serious problem encountered in the application of affinity chromatography to the estrogen receptor system has been the small but definite leakage or bleeding of the immobilized estrogen from the matrix. The leakage of the ligand seems to result, at least in part, from the cleavage of the initial chemical bonds formed during the CNBr activation step. This problem of ligand leakage has recently been approached by two different methods. First, the substituted agarose is diluted with an optimal quantity of unsubstituted agarose, thus lowering the net amount of ligand on the adsorbent and consequently the resulting leakage. Table II illustrates this approach with data obtained with ganglioside–agarose derivatives used in the purification of cholera toxin (Cuatrecasas *et al.*, 1973). The second approach involves the use of polyfunctional reagents like the branched-chain copolymer of Lys and Ala (Figure 4). Such polymers result in multipoint attachment when reacted under appropriate conditions with CNBr-activated agarose. A ligand attached to agarose via such a polyfunctional polymer arm has greater stability, since the statistical chances of release of a ligand from the matrix will diminish in geometric proportion as the number of points of attachment increases. Besides enhanced stabilization of the ligand, the polyfunctional reagents provide large numbers of functional groups (α-NH_2 groups, in the example cited) for potential ligand substitution sites, a favorable microenvironment around the ligand, and a long extension arm between the ligand and the matrix (Parikh *et al.*, 1974). Bovine serum albumin has also been used as a polyfunctional extension arm during affinity chromatography of cholera toxin (Cuatrecasas *et al.*, 1973) and estrogen receptors (Parikh *et al.*, 1974*b*). Figure

TABLE II

Chromatography of ^{125}I-Labeled Cholera Toxin on Various Affinity Adsorbents[a]

Agarose derivative	% of ^{125}I-labeled toxin		Ganglioside in elution, % inhibition
	in breakthrough	in elution	
Unsubstituted agarose	93	3	
Diaminodipropylamino-agarose	91	7	
A-DADA-gang			
Undiluted	21[b]	74	48
1:10	27[b]	70	20
1:50	48	41	0
A-PLL-Ala-gang			
Undiluted	17[b]	80	5
1:10	19[b]	78	0
1:50	20	76	0
1:200	52	42	0
A-NatAlb-gang			
Undiluted	18[b]	78	10
1:10	20[b]	81	0
1:50	21[b]	75	0
1:200	38	52	0
1:600	68	30	0

[a] Columns (Pasteur pipets) containing 1 ml of the specified gel were washed with 10 ml each of 50% methanol, 7 M guanidine·HCl, and Krebs–Ringer–bicarbonate buffer. ^{125}I-labeled cholera toxin (2.1 × 10^5 cpm), in 0.5 ml of Krebs–Ringer–bicarbonate buffer, containing 0.1% albumin, was applied to each column. The columns were washed, eluted with 5 M guanidine·HCl in 50 mM Tris·HCl (pH 7.4), containing 0.1% albumin. In some cases the adsorbents were diluted serially with unsubstituted agarose before use. The eluted samples were examined for the presence of free gangliosides. The presence of ganglioside is expressed by the percent inhibition of binding. None of the breakthrough samples contained free gangliosides.
[b] More than 95% of this radioactivity corresponds to inactive toxin present in the samples which were applied to the columns. Data adapted from P. Cuatrecasas, I. Parikh, and M. Hollenberg, *Biochemistry* **12**: 4253 (1973).

5 compares the purification of estrogen receptors obtained with three different extension arms (spacers) under otherwise identical experimental conditions. The relatively lower efficiency of albumin as an interposing macromolecular extension in this case is attributed to its strong binding to the estradiol which is present in the eluting buffer. Because the estrogen receptor proteins are sensitive to minor changes in pH or ionic strength of the buffers, the elution from the affinity column is done with buffers containing the ligand, estradiol (Figure 6). As the spontaneous dissociation of the receptor from the agarose-immobilized ligand is a slow process, too large an excess of free ligand in the eluting buffer will not effectively enhance the efficacy of elution. The key

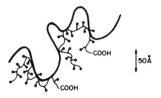

DENATURED ALBUMIN-AGAROSE

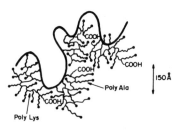

POLY-L-LYSINE-POLY-DL-ALANINE-AGAROSE

FIGURE 4. Branched chain copolymer of Lys and Ala. Average molecular weight 37,500. An average chain of 15 Ala units are attached to each of the ε-NH_2 groups of the Lys backbone. Considering a random coil structure, the average size of the macromolecule is approximated to 150 Å.

kinetic parameter in the elution is the off rate, k_{-1}, which is highly dependent on temperature. Thus, this type of ligand exchange can be accelerated very substantially by increasing the temperature. Because the exchange process is also time-dependent, ligand elution is often performed as a batchwise procedure, allowing sufficiently prolonged time of contact to achieve complete exchange.

Insulin Receptors.   As in the case of estrogen receptors, insulin receptors pose special problems due to the extraordinarily small quantities available in biological tissues. Furthermore, the hydrophobic character of this receptor, which requires the presence of nonionic detergents to maintain solubility, presents additional difficulties during affinity chromatography of these proteins (Cuatrecasas, 1972). It is estimated that to achieve complete purification of this receptor requires 400,000-fold purification of the crude liver homogenate, in contrast to the purification of acetylcholine receptors (also membrane bound), which require only 300- to 1000-fold purification, depending on the source. Conventional agarose derivatives of insulin, coupled through residues B1 or B29 and containing 20- to 40-Å extension arms, are effective in retaining the solubilized insulin receptor. The macromolecular insulin–agarose adsorbents are found to be much more effective in the affinity chromatography of this receptor (Cuatrecasas and Parikh, 1974). Since solubilized

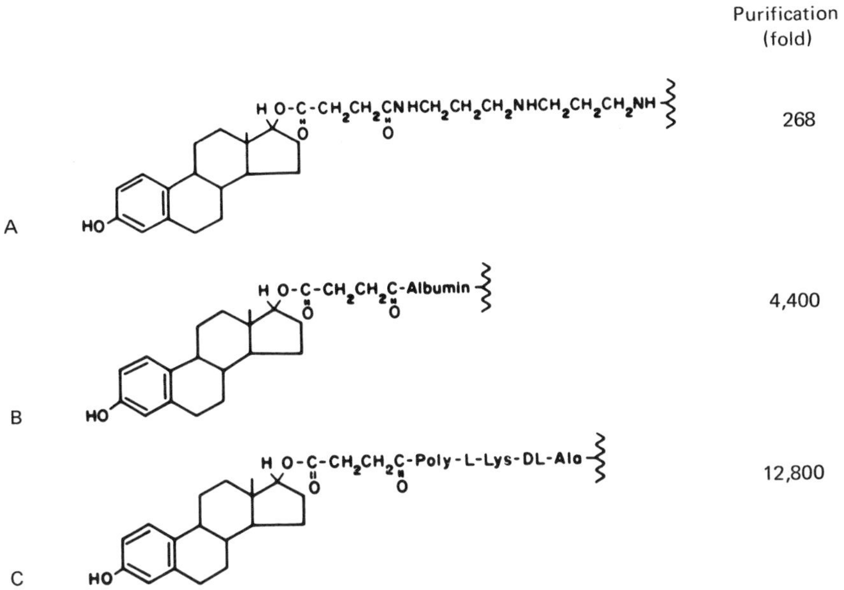

Purification
(fold)

A

268

B

4,400

C

12,800

FIGURE 5. Comparative efficiencies of three agarose adsorbents used in the isolation of estrogen receptor from immature calf uterus cytosol. Columns (0.9 × 9.5 cm) containing various affinity adsorbents having approximately equal substitution of the ligand were used. The elution of the receptor proteins was achieved in batchwise manner by incubating the gel in buffer containing 10 μM [^{3}H]estradiol (4 Ci/mmole) for 15 min at 30°. The eluted receptor was tested for binding on Sephadex G-25 before and after heating for 5 min at 65°. Data adapted from Sica *et al.* (1973).

insulin receptors appear to be glycoproteins which are capable of interacting with certain plant lectins, adsorbents containing such lectins have also been successfully explored for purification. Elution of the receptor proteins from such lectin–agarose columns can be achieved under very mild condition with buffers containing one of the lectin-specific simple sugars (Cuatrecasas, 1973).

Transport Proteins. Fannin and Diedrich (1973) used the technique of affinity chromatography to isolate the putative, much disputed glucose-carrier protein from the rabbit small-intestinal epithelium cells by using phloretin coupled to agarose as the affinity adsorbent. Phloretin is a phenolic polycyclic glucoside which is known to specifically inhibit active sugar transport.

Plant Auxin Acceptor Proteins. Affinity columns containing an immobilized plant auxin, namely, 2,4-dichlorophenoxyacetic acid, were recently

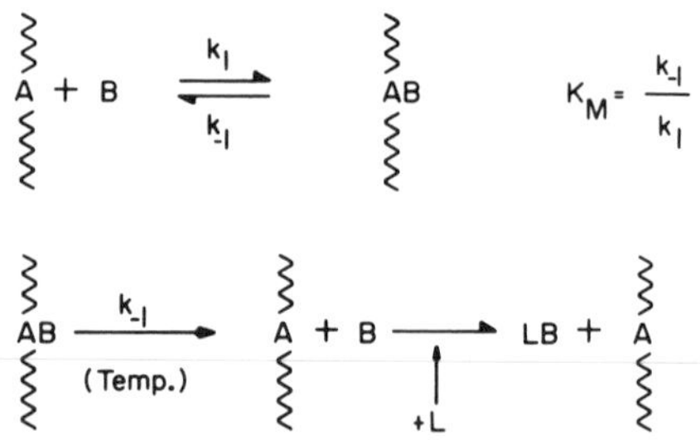

$$A + B \underset{k_{-1}}{\overset{k_1}{\rightleftharpoons}} AB \qquad K_M = \frac{k_{-1}}{k_1}$$

$$AB \xrightarrow[(\text{Temp.})]{k_{-1}} A + B \longrightarrow LB + A$$
$$\uparrow +L$$

FIGURE 6. Ligand elution. Elution of a protein from an affinity column is often performed under milder conditions with buffers containing a specific ligand. The eluting ligand may or may not be the same as the one which is immobilized on the matrix. Ligand elution is often performed in batchwise manner. Increased temperature and longer incubation times may help to achieve maximum exchange between matrix-bound ligand (A) and the free ligand (L).

used (Venis, 1971) to isolate the auxin acceptor proteins from crude extracts of pea and corn shoots. These proteins, which stimulate *E. coli* DNA-dependent RNA polymerase, were eluted from the affinity column with 2 mM KOH. Enzyme stimulation by the partially purified preparation was increased twofold. The extent of purification could not be estimated because the amount of acceptor protein present in the starting crude homogenate could not be determined, probably because of the presence of some inhibitory proteins. By virtue of the very small quantities of acceptor protein isolated, binding studies of radiolabeled auxin with the purified preparation were not possible.

Gal Repressor. The repressor of the galactose operon of *E. coli* has been partially purified by affinity chromatography, and it has been identified as a protein (Parks *et al.*, 1971). The galactose binding properties of the gal repressor were used as the basis for its purification. The ligand, *p*-aminophenyl-β-D-thiogalactoside, linked to agarose, provided a single-step convenient method of separating the gal repressor from other DNA-binding proteins. β-Galactosidase activity was separated from the gal repressor by differential elution of the affinity column, resulting in greater than 300-fold purification of the latter protein. A major contaminant in the purified gal repressor was found to be lac repressor protein. Similarly, copurification of lac repressor and β-galactosidase by affinity chromatography was reported using the same ligand linked to bovine gamma globulin (Tomino and Paigen, 1970).

ara–DNA Operon. The purification of regulatory proteins present similar kinds of problems as those discussed for insulin and estrogen receptors. They are present in very low concentrations in the cell, they are often unstable, and they are not easily assayed. The L-arabinose operon seems to exist in two functionally active conformational states, either as an activator or repressor

of gene expression. L-Arabinose acts as an allosteric ligand. *p*-Aminophenyl-
β-D-fucopyranoside, selected as specific ligand, was attached covalently to
agarose through a small extension arm by the standard cyanogen bromide
method (Wilcox *et al.*, 1971). The selection of D-fucose as an affinity ligand
was based on the observation that it interacts with the L-arabinose operon
in vivo, and is not metabolized by the *E. coli* strain lacking the enzyme
L-arabinose isomerase. The ligand also prevents the induction of the
L-arabinose operon in the presence of L-arabinose. The crude *E. coli* extracts
containing ara operon are loaded on the affinity column at pH 7 and eluted,
after washing the contaminating proteins, with a pH 10 buffer containing
1 mM dithiothreitol. The overall purification appears to have been between
300- and 1000-fold (Wilcox *et al.*, 1971).

   Covalent Affinity Chromatography. Certain drugs exert their pharma-
cological effects by inhibiting specific metabolic enzymes through covalent
reaction. Such drugs, or substrates which in their normal process of
catalysis participate in covalent enzyme–substrate intermediates (even
though transiently), may be used for a special type of "covalent" affinity
chromatography. For example, such approaches have been used successfully
in the isolation and purification of D-alanine carboxypepdidase (Blumberg
and Strominger, 1972) and acetylcholine esterase (Ashani and Wilson, 1972).

   An affinity adsorbent consisting of agarose-bound 6-aminopenicillinic
acid binds five different detergent-solubilized cell wall proteins of *Bacillus
subtilis*. The adsorbed proteins, which are bound covalently to the penicillinic
acid, can be eluted quantitatively from the affinity column with neutral
solutions of hydroxylamine hydrochloride, since this nucleophile cleaves the
penicilloyl–enzyme bond. If the detergent-solubilized proteins are pretreated
with low concentrations of cephalothin (a penicillin analog), only one protein,
D-alanine carboxypeptidase, is retarded on the 6-aminopenicillinic acid–
agarose affinity column. The carboxypeptidase, which consists of up to 1%
of the total cell wall protein, can be eluted from the affinity column with
0.8 M neutral hydroxylamine in 50% overall yield (Figure 7).

   Acetylcholine esterase and some other esterases are inhibited "irreversibly" by a variety of organophosphates containing a good leaving group,
since they form covalent phosphoryl–enzyme complexes devoid of enzymic
activity. The activity of the inhibited esterase can be restored with specific
nucleophiles such as 2-PAM, TMB$_4$, or even hydroxylamine. Affinity columns
of agarose–*p*-nitrophenyl-2-ethylamino methylphosphonate adsorb acetylcholine esterase through covalent reaction; *p*-nitrophenol is the leaving group.
Ashani and Wilson (1972) isolated acetylcholine esterase from electric eel
using this type of covalent affinity-chromatographic technique, as illustrated
in Figure 8. The adsorbed enzyme, which is devoid of enzymic activity, is
eluted in active form with buffers containing a strong nucleophile, such as

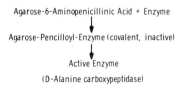

FIGURE 7. Covalent affinity chromatography of penicillin-binding components. The 6-aminopenicillinic acid was coupled to succinyl-diaminodipropylamino-agarose with a water-soluble carbodiimide. Detergent-solubilized preparation of *B. subtilis* membranes, pretreated with cephalothin, was applied to the column. The enzyme, D-alanine carboxypeptidase (99% pure), was eluted with 0.8 M neutral hydroxylamine hydrochloride. In the absence of cephalothin pretreatment of the solubilized homogenate the affinity column copurifies four to five different penicillin-binding proteins (Blumberg and Strominger, 1972).

10mM 2PAM at pH 8.0, in about 40% yield. An additional 40% to 50% of the enzyme is recovered if the adsorbent is incubated with the eluting buffer for another 48 h at room temperature. Although α-chymotrypsin is also purified on this affinity column, not all serine esterases may be adsorbed on such adsorbents because of their variable reactivity to the ligand. The operative capacity for such adsorbents tends to be remarkably high. Thus, up to 2.5 mg of α-chymotrypsin may be adsorbed per milliliter of packed adsorbent when the ligand substitution is 0.2 μmoles per milliliter of packed gel. As with aminopenicillinic acid–agarose, the affinity adsorbent for acetylcholine esterase

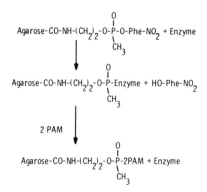

FIGURE 8. Covalent affinity chromatography of acetylcholinesterase. Agarose adsorbents, prepared by attaching 2-aminoethyl-*p*-nitrophenyl methylphosphonate, reacts covalently with acetylcholine esterase by displacement reaction. *p*-Nitrophenol is the leaving group. The immobilized enzyme, devoid of enzymic activity, is eluted in active form with buffers containing one of the strong nucleophiles such as 2-PAM, TMB₄, or even neutral hydroxylamine hydrochloride (Ashani and Wilson, 1972).

cannot be regenerated once it has been used in the affinity-chromatographic purifications.

Covalent affinity chromatography has also been used in the purification of thiol-containing proteins such as papain. In this technique, the ligand consists of a low-molecular-weight disulfide-containing compound which is attached covalently to a matrix. The SH-protein undergoes a thiol–disulfide interchange reaction, resulting in the formation of a matrix-disulfide-protein adduct. The covalently adsorbed protein can then be eluted from the affinity column with buffers containing mercuric chloride (Sluytermann and Wijdenes, 1970), β-mercaptoethanol, or dithiothreitol (Brocklehurst and Little, 1973).

Hydrophobic Affinity Chromatography. Although hydrophobic interactions between ligands and biological macromolecules can rarely be described as being "specific," they can be exploited in some cases to purify proteins. This empirical method is based on the observation that a hydrocarbon chain of optimum length, when attached covalently to a solid matrix, may adsorb one protein in preference to another. Elutions of such adsorbed proteins are realized by changing (i.e., increasing) the ionic strength and other parameters of the eluting buffer (Figure 9). In general, three main types of hydrocarbon chains are used (Shaltiel, 1974; Hofstee, 1973). The nonionic amphiphilic ligands are generally prepared by coupling a mono- or a dihydrazide to cyanogen-bromide-activated agarose. The hydrazide functions having very low pK (4 to 4.5) are not charged during the use of common buffers. The ionic-hydrophobic matrices may carry one or two positive charges depending on whether they are prepared by coupling a monoamine or a diamine to the

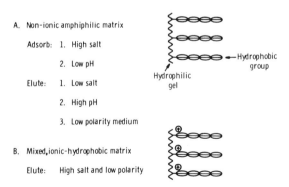

A. Non-ionic amphiphilic matrix

    Adsorb: 1. High salt

              2. Low pH

    Elute: 1. Low salt

              2. High pH

              3. Low polarity medium

B. Mixed, ionic-hydrophobic matrix

    Elute: High salt and low polarity

FIGURE 9. Hydrophobic affinity chromatography. The nonionic amphiphilic matrices are often prepared by coupling straight-chain aliphatic acid hydrazides to cyanogen-bromide-activated agarose. The entire adsorbent presumably does not carry any charge above pH 4. The mixed, ionic–hydrophobic adsorbents possess at least one ionic group and a hydrophobic alkyl chain.

cyanogen-bromide-activated agarose. The general elution strategies are summarized in Figure 9. Hydrophobic affinity chromatography has been discussed in depth in the second volume of this series by Dr. S. Hjertén and Dr. B. Hofstee.

Group-Specific Affinity Adsorbents. This is a class of affinity adsorbents possessing a general low-molecular-weight ligand (e.g., coenzymes, cofactors), which is recognized rather specifically by a group or series of biological macromolecules rather than by one or a very limited number (depending on the substrate specificity). The affinity of the ligand toward the individual macromolecules within a group may vary only slightly, but the participation of other substrates or variations in experimental conditions may permit affinity-chromatographic separation of the macromolecules under carefully selected conditions. The resolving power of such group-specific adsorbents lies primarily in the properly designed desorption technique. The elution is based on the differential affinities of the macromolecules toward the immobilized ligand, or in the other specific components added to the buffer, such as substrates, cofactors, or competitive inhibitors. A novel kind of group-specific affinity adsorbent was recently prepared by covalent immobilization of Cibacron Blue F3GA dye (the dye which is also present in the commercial Blue Dextran) to agarose. The adsorbent (Figure 10) has been successfully used in the purification of phosphofructokinase (Bohme *et al.*, 1972) and other kinases, as well as dehydrogenases (Easterday and Easterday, 1974). Although the interaction between the blue ligand and the enzymes is definitely not active site specific, the elution methods are made very specific through inclusion of a substrate, cofactor, or an inhibitor of the enzyme in the elution buffers. Included among other types of group-specific affinity adsorbents is 5'-AMP-

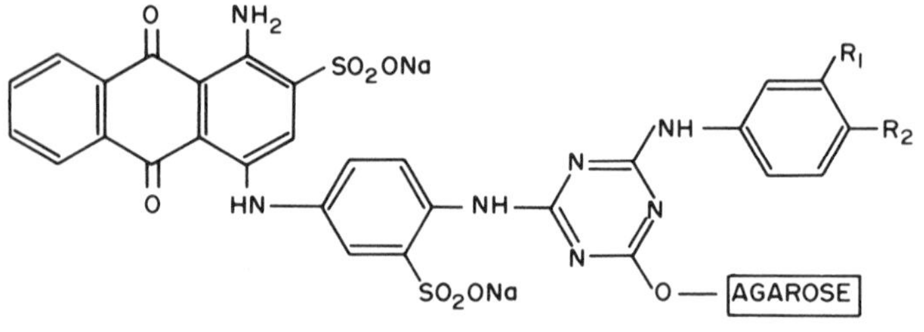

CIBACRON BLUE F3GA
(BLUE DEXTRAN)

FIGURE 10. Group-specific affinity ligand. Cibacron Blue F3GA. The substitution of the blue dye on agarose is generally between 0.75 and 2.1 μmoles per milliliter of packed gel. Approximately 8.0 μmoles of the dye can be substituted per milliliter of Sephadex G.50.

agarose (Dean *et al.*, 1974; Mosbach *et al.*, 1972), which is a general ligand for a group of enzymes such as kinases and dehydrogenases which have substrates or cofactors containing an adenine residue. Agarose derivatives containing immobilized concanavalin A (Con A) can also be considered as group-specific general affinity ligands. The specificity of Con A–agarose is restricted to all polysaccharides and glycoproteins which contain α-D-mannosyl or α-D-glucosyl residues.

Quantitative Aspects. Although a good amount of knowledge has been accumulated about the general applicability of affinity chromatography, few theoretical guidelines and quantitative analyses based on the available physicochemical parameters are available. Simple kinetic and equilibrium models for the affinity adsorption and desorption events have recently been attempted (Graves and Wu, 1974). Mathematical analysis of this type should ultimately lead to a better understanding of affinity chromatography results and should allow a more efficient application of this technique to future problems. Through simple kinetic analysis of the effects of various sugars on the elution behavior of the A-protein of lactose synthetase from α-lactalbumin affinity columns, it was possible to calculate association constants of the protein and the eluting ligand (Andrews *et al.*, 1973). Recently, Wankat (1974) has developed a mathematical theory which evaluates the loading and elution events of an affinity column in terms of steric hindrance of ligand sites and enzyme–ligand and enzyme–enzyme interactions. Similarly, affinity chromatography has been utilized in quantitating the affinity of proteins to both matrix bound ligand as well as the competing ligand present in the elution buffers (Dunn and Chaiken, 1974).

Immobilized Polynucleotides. The use of immobilized polynucleotides, nucleic acids, and nucleotide phosphates has been ever increasing in affinity-chromatographic procedures. In certain cases such immobilized polynucleotides have been ingeniously used to study the mechanism of action of certain enzymes. The methods generally used for the immobilization of polynucleotides have been summarized in Table III. The immobilization and use of nucleotide phosphates is a vast field in itself and shall not be reviewed here. The coupling of single-stranded DNA and RNA to cyanogen-bromide-activated agarose is one of the more widely used methods of immobilization (Poonian *et al.*, 1971). The primary amino groups of some of the nucleotide bases are presumably involved in this covalent attachment reaction. It has been suggested, although it appears to be less likely, that if the coupling reaction with cyanogen-bromide-activated agarose is performed at pH 6, the polynucleotides may be immobilized through the terminal phosphate group (Wagner *et al.*, 1971). It is conceivable that coupling of denatured DNA or RNA by CNBr-activation method results in multipoint attachments. Periodate oxidation of the 3′-terminal *cis*-hydroxyl groups in RNA and polyribonucleo-

TABLE III
Immobilization of Polynucleosides
and Nucleic Acids

1. Cyanogen-bromide-activated agarose
2. Periodate oxidation of 3'-terminals in RNA
3. Complexing with organoboryl group
4. UV irradiation
5. Condensation with phosphocellulose

tides, and coupling of the resulting dialdehydes to amino- or hydrazido-agarose, has been reported (Gilham, 1974; Robberson and Davison, 1972). The 3'-terminal of RNA also offers an additional alternative for immobilization, namely, through complexing with matrix-bound organoboryl groups (Rosenberg et al., 1972). It is interesting to note that such complexation would occur only between pH 8 and 9. However, at pH6 the complex breaks down, allowing the recovery of the polyribonucleotide and possible reuse of the organoboryl matrix. Ultraviolet irradiation of the solutions of DNA, RNA, or polynucleotides in the presence of an inert matrix offers a good (but chemically less well understood) means of immobilization (Litman, 1968). The substitution takes place through multipoint attachments. Bautz and Hall (1962) were able to couple phage DNA to acetylated phosphocellulose by using standard carbodiimide procedures. The affinity adsorbents were used to isolate T4-specific RNA based upon the principle of RNA–DNA hybrid formation.

Immobilized polynucleotides have also been very widely used in the purification of various enzymes involved in the synthesis or regulation of nucleic acids. Oligo(dt)-cellulose can be used to specifically isolate A-rich polynucleotides from the total cellular RNA on the principle of base pairing. The method is quite general and can be useful in the isolation of mRNA species containing particular oligonucleotide sequences by selective adsorbents prepared by immobilization of a complementary base sequence. The immobilized polynucleotides have also found a novel application in the study of the mechanism of action of certain important biological processes. Poly(I)-agarose and its base-paired complex formed with poly(C) have been used as probes for the study of the mechanism of induction of host resistance to viral infection (Wagner et al., 1971).

Use of Plant Lectins. Plant lectins are a remarkable group of complex proteins which agglutinate erythrocytes and have a wide range of specificity toward mono-, di-, and oligosaccharides in particular sequence. The property of plant lectins to recognize and bind to specific sugar residues has led to their application in the purification of various glycoproteins. Concanavalin A, the most widely used plant lectin, is a metalloprotein containing Mn^{++} and requires Ca^{++} for binding to its specific sugars, α-D-mannopyranosides and α-D glucopyranosides. Any glycoprotein containing an α-D-mannopyranosyl or an α-D-glucopyranosyl residue with their C-3, C-4, and C-6 hydroxyl group free will bind to agarose to which Con A has been covalently immobilized. Con A–agarose adsorbents have been widely used in the purification of various glycoproteins such as rhodopsin, interferon, and others. Table IV summarizes some examples of plant lectin–agarose adsorbents and their applications. Only a few of the typical examples are discussed here. The plant lectins themselves have also been successfully purified on affinity adsorbents containing insolubilized complementary monosaccharides or glycoproteins (Table V).

Rhodopsin. The glycoprotein character of this protein of the bovine retinal disc membrane was utilized for purification with columns containing agarose-immobilized Con A (Steineman and Stryer, 1973). The coupling of Con A to cyanogen-bromide activated agarose is performed at pH 6.8, presumably to achieve attachments at the α-NH_2 groups of the protein. The detergent solubilized preparation of bovine retinal disc membranes was used in the affinity-chromatographic procedure. Almost all of the rhodopsin present in the solubilized homogenate is adsorbed on the Con A–agarose column. The adsorbed rhopsin is eluted quantitatively with 0.1 M D-glucose. It is significant to note that the selectivity and capacity of the Con A–agarose is maintained

TABLE IV

Application of Various Plant Lectins in the Purification of Certain Glycoproteins

Protein	Lectin	Reference
Rhodopsin	Con A	Steineman and Stryer, 1973
HCG, LH, and FSH	Con A	Dufau et al., 1972
Teichoic acid	Con A	Doyle et al., 1973
α-Antitrypsin	Con A	Murthy and Hercz, 1973
Virus glycoproteins	*Lens culinaris*	Hayman et al., 1973
Interferon	Con A	Davey et al., 1974
	Phaseolus vulgaris	Dorner et al., 1973
Blood group substances	*Vicia cracca*	Kristiansen, 1974; Dowson et al., 1974
Lectin receptors	*Lens culinaris*	Hayman and Crumpton, 1972; Findlay, 1974; Adair and Kornfeld, 1974
	Con A	Allan et al., 1972
Insulin receptor	Con A and WGA	Cuatrecasas, 1973
Dopamin-β-hydroxylase	Con A	Rush et al., 1974
Liver-β-galactosidase	Con A and WGA	Norden and O'Brien, 1974
Glycogen synthetase	Con A	Solling and Wang, 1973
Arylsulfatase-A	Con A	Ahmed et al., 1973

TABLE V

Affinity Chromatographic Purification of Certain Plant Lectins

Lectin	Specific sugar	Affinity ligand	Matrix	Reference
Concanavalin A (jack beans; Con A)	α-D-Mannose and α-D-glucose	Dextran	Sephadex	Agraval and Goldstein, 1967
Wheat germ agglutinin (WGA)	N-acetyl-D-glucosamine (NAG)	Ovomucoid	Sephadex	Levine et al., 1972
		Chitin	—	Bloch and Burger, 1974
		NAG	Agarose	Lotan et al., 1973
Sophora japonica	D-Galactose	Gastric mucin	—	Poretz et al., 1974
Ricinus communis (castor beans)	D-Galactose	Dextran	Sephadex	Nicolson et al., 1974
Soybean agglutinin	N-acetyl-galactosamine and D-galactose	Galactosylamine	Agarose	Gordon et al., 1972
Phaseolus vulgaris	N-acetyl-galactosamine	Thyroglobulin	Agarose	Matsumoto and Osawa, 1972

even in the presence of 1.4% detergent, cethyltrimethylammonium bromide, in the buffers. As rhodopsin contains a carbohydrate moiety consisting of equal quantities of N-acetylglucosamine and mannose, it is probable that adsorbents prepared from wheat germ agglutinin may also bind to the retinal disc membranes.

Glycoprotein Hormones. Con A–agarose adsorbents were used by Dufau *et al.* (1972) to purify three glycoprotein hormones: human chorionic gonadotrophin (HCG), human luteinizing hormone (LH), and follicle-stimulating hormone (FSH). Human chorionic gonadotrophin hormone and its subunits were adsorbed on affinity columns containing Con A–agarose and could be eluted with 0.2 M α-methyl-D-glucopyranoside in about 85% yield. Chromatography of asialo-HCG on a Con A affinity column showed a distinctly enhanced retardation of the asialo hormone as compared to the native hormone. Furthermore, the asialo-agalacto-HCG was even more strongly adsorbed on the Con A–agarose column since it could not be eluted with 1 M α-methyl-D-glucopyranoside or with 1.0 M EDTA. These experiments suggest that the interaction of the glycoprotein with the plant lectin is favored by removal of the intervening sugar residues which expose the manosyl moiety on the surface. Prior combination of ^{125}I-labeled HCG with specific antibody did not prevent its binding to Con A–agarose. The antibody–antigen complex was subsequently eluted with α-methyl-D-glucopyranoside in high yield.

Interferon. Attempts to purify the antiviral glycoprotein interferon are numerous, but none has been sufficiently successful in providing a reasonable state of purity. Recently Doyle *et al.* (1973) showed that human interferon can selectively bind to Con A–agarose. A substantial purification of interferon was achieved when the affinity column was eluted with buffer containing 0.1 M α-methyl-D-mannopyranoside and 50% ethylene glycol. Dorner *et al.* (1973) have used a lectin from red kidney beans (*Phaseolus vulgaris*) as an affinity ligand for the purification of desialated human interferon. The advantage of this plant lectin seems to be its ability to recognize a relatively complex oligosaccharide sequence, galactose $\rightarrow$ N-acetylglucosamine $\rightarrow$ mannose. The adsorbed interferon was eluted with a solution of a glycoprotein fragment of human erythrocytes of known structure. This glycopeptide contains a branched carbohydrate structure that seems to compete successfully with the asialo-interferon for the binding sites of the plant lectin.

REFERENCES

Adair, W. L. and Kornfield, S. (1974) *J. Biol. Chem.* **249**: 4696.
Agraval, B. B. L. and Goldstein, I. J. (1967) *Biochim. Biophys. Acta* **147**: 262.

Ahmed, A., Bishayee, S., and Bachhawat, B. K. (1973) *Biochem. Biophys. Res. Commun.* **53**: 730.

Allan, D. Auger, J., and Crumpton, M. J. (1972) *Nature New Biol.* **236**: 23.

Andrews, P. (1970) *FEBS Lett.* **9**: 297.

Andrews, P., Kitchen, B. J., and Winzor, D. J. (1973) *Biochem. J.* **135**: 897.

Ashani, Y. and Wilson, I. B. (1972) *Biochim. Biophys. Acta* **276**: 317.

Bautz, E. K. F. and Hall, B. D. (1962) *Proc. Nat. Acad. Sci. USA* **48**: 400.

Barker, R., Olsen, K. W., Shaper, J. H., and Hill, R. L. (1972) *J. Biol. Chem.* **247**: 7135.

Bloch, R. and Burger, M. M. (1974) *Biochem. Biophys. Res. Commun.* **58**: 13.

Blumberg, P. M. and Strominger, J. L. (1972) *Proc. Nat. Acad. Sci. USA* **69**: 3751.

Bohme, H. J., Kopperschlager, G., Schulz, J., and Hofmann, E. (1972) *J. Chromatogr.* **69**: 209.

Brocklehurst, K. and Little, G. (1973) *Biochem. J.* **133**: 67.

Chauvet, J. and Acher, R. (1973) *Biochimie* **55**: 1323.

Cuatrecasas, P. (1970) *J. Biol. Chem.* **245**: 3059.

Cuatrecasas, P. (1972) *Proc. Nat. Acad. Sci. USA* **69**: 1277.

Cuatrecasas, P. (1973) *J. Biol. Chem.* **248**: 3528.

Cuatrecasas, P. and Anfinsen, C. B. (1971) *Ann. Rev. Biochem.* **40**: 259.

Cuatrecasas, P. and Illiano, G. (1971) *Biochem. Biophys. Res. Commun.* **44**: 178.

Cuatrecasas, P. and Parikh, I. (1974) in *Methods in Enzymology*, Vol. 34 (Jacoby and Wilchek, eds., pp. 653–670, Academic Press, New York.

Cuatrecasas, P., Wilchek, M., and Anfinsen, C. B. (1968) *Proc. Nat. Acad. Sci. USA* **61**: 636.

Cuatrecasas, P., Parikh, I., and Hollenberg, M. D. (1973) *Biochemistry* **12**: 4253.

Davey, M. W., Huang, J. W., Sulkowski, E., and Carter, W. A. (1974) *J. Biol. Chem.* **249**: 6354.

Dean, P. D. G., Craven, D. B., Harvey, M. J., and Lowe, C. R. (1974) in *Immobilized Biochemicals and Affinity Chromatography* (Dunlap, ed.) pp. 99–121, Plenum Press, New York.

Dorner, F., Scriba, M., and Weil, R. (1973) *Proc. Nat. Acad. Sci. USA* **70**: 1981.

Dowson, J. R., Silver, J., Sheppard, L. B., and Amos, D. B. (1974) *J. Immunol.* **112**: 1190.

Doyle, R. J., Birdsell, D. C., and Young, F. E. (1973) *Prep. Biochem.* **3**: 13.

Dufau, M. L., Tsuruhara, T., and Catt, K. J. (1972) *Biochem. Biophys. Acta* **278**: 281.

Dunn, B. M. and Chaiken, I. M. (1974) *Proc. Nat. Acad. Sci. USA* **71**: 2382.

Easterday, R. L. and Easterday, I. M. (1974) in *Immobilized Biochemicals and Affinity Chromatography* (Dunlap, ed.) pp. 123–133, Plenum Press, New York.

Fannin, F. F. and Diedrich, D. F. (1973) *Arch. Biochem. Biophys.* **158**: 919.

Feinstein, G. (1970) *Biochim. Biophys. Acta* **214**: 224.

Feldman, K., Ziesel, H., and Helmreich, E. (1972) *Proc. Nat. Acad. Sci. USA* **69**: 2278.

Findlay, J. B. C. (1974) *J. Biol. Chem.* **249**: 4398.

Foster, R. J. and Ryan, C. R. (1974) *Biochemistry* **13**: 132.

Fressinaud, P. H., Corval, P., Frenaz, J. P., and Menard, J. (1973) *Biochim. Biophys. Acta* **317**: 572.

Gilham, P. T. (1974) in *Immobilized Biochemicals and Affinity Chromatography* (Dunlap, ed.) pp. 173–185, Plenum Press, New York.

Gordon, J. A. Blumberg, S., Lis, H., and Sharon, N. (1972) *FEBS Lett.* **24**: 193.

Graves, D. J. and Wu, Yun-Tai (1974) in *Methods in Enzymology*, Vol. 34 (Jacoby and Wilchek, eds.) pp. 140–163, Academic Press, New York.

Green, M. (1973) *Biochem. J.* **133**: 698.

Green, M. and Torns, E. J. (1973) *Biochem. J.* **133**: 687.

Hayman, M. J. and Crumpton, M. J. (1972) *Biochem. Biophys. Res. Commun.* **47**: 923.
Hayman, M. J., Skehel, J. J., and Crumpton, M. J. (1973) *FEBS Lett.* **29**: 185.
Helman, M. and Givol, D. (1971) *Biochem. J.* **125**: 971.
Hofstee, B. H. J. (1973) *Anal. Biochem.* **52**: 430.
Javid, J. and Liang, J. (1973) *J. Lab. Clin. Med.* **82**: 991.
Kato, I. and Anfinsen, C. B. (1969) *J. Biol. Chem.* **244**: 1004.
Kenyon, A. J., Gander, J. E., Lopez, C., and Good, R. A. (1973) *Science* **179**: 187.
Klein, M. and Mihaesco, C. (1973) *Biochem. Biophys. Res. Commun.* **52**: 774.
Konda, H., Hayashi, H., and Mikashi, K. (1972) *J. Biochem.* **72**: 759.
Kristiansen, T. (1974) *Biochim. Biophys. Acta* **338**: 246.
Landman, A. D. and Dakshinamurti, K. (1973) *Anal. Biochem.* **56**: 191.
LeVine, D., Kaplan, M. J., and Greenaway, P. J. (1972) *Biochem. J.* **129**: 847.
Litman, R. M. 1968. *J. Biol. Chem.* **243**: 6222.
Lornitzo, F. A., Qureshi, A. A., and Porter, J. W. (1974) *J. Biol. Chem.* **249**: 1654.
Lotan, R., Gussin, A. E. S., Lis, H., and Sharon, N. (1973) *Biochem. Biophys. Res. Commun.* **52**: 656.
Mosbach, K., Guilford, H., Ohlsson, R., and Scott, M. (1972) *Biochem. J.* **127**: 625.
Motsumoto, I. and Osawa, T. (1972) *Biochem. Biophys. Res. Commun.* **46**: 1810.
Murthy, R. J. and Hercz, A. (1973) *FEBS Lett.* **32**: 243.
Nicolson, G. L., Blaustein, J., and Etzler, M. E. (1974) *Biochemistry* **13**: 196.
Norden, A. G. W. and O'Brien, J. S. (1974) *Biochem. Biophys. Res. Commun.* **56**: 193.
Ohlsson, K. and Tegner, H. (1973) *Biochim. Biophys. Acta* **317**: 328.
Ontjes, D. A. and Anfinsen, C. B. (1969) *J. Biol. Chem.* **244**: 6316.
Parikh, I., Corley, L., and Anfinsen, C. B. (1971) *J. Biol. Chem.* **246**: 7392.
Parikh, I., March, S., and Cuatrecasas, P. (1974) in *Methods in Enzymology*, Vol. 34 (Jacoby and Wilchek, eds.) pp. 77–102, Academic Press, New York.
Parikh, I., Sica, V., Nola, E., Puca, G. A., and Cuatrecasas, P. (1974) in *Methods in Enzymology*, Vol. 34 (Jacoby and Wilchek, eds.) pp. 670–688, Academic Press, New York.
Parks, J. S., Gottesman, M., Shimada, K., Weisberg, R. A., Perlman, R. L., and Pastan, I. (1971) *Proc. Nat. Acad. Sci. USA* **68**: 1891.
Pass, L., Zimmer, T. L., and Loland, S. G. (1973) *Eur. J. Biochem.* **40**: 43.
Poonian, M. S., Schlabach, A. J., and Weissbach, A. (1971) *Biochemistry* **10**: 424.
Poretz, R. D., Riss, H., Timberlake, J. W., and Chien, S. (1974) *Biochemistry* **13**: 250.
Pradelles, P., Margot, J. L., and Fromageot, P. (1972) *FEBS Lett.* **26**: 189.
Robberson, D. L. and Davison, N. (1972) *Biochemistry* **11**: 533.
Robinson, N. C., Tye, R. W., Neurath H., and Walsh, K. A. (1971) *Biochemistry* **14**: 2743.
Rosenberg, M., Wiebers, D. L., and Gilham, P. T. (1972) *Biochemistry* **11**: 3623.
Rush, R. A., Thomas, P. E., Kindler, S. H., and Udenfriend, S. (1974) *Biochem. Biophys. Res. Commun.* **57**: 1301.
Schmit, J. C., Artz, S. W., and Zalkin, H. (1970) *J. Biol. Chem.* **245**: 4019.
Shaltiel, S., 1974. in *Methods in Enzymology* (Jacoby and Wilchek, eds.) pp. 126–140, Academic Press, New York.
Sica, V., Parikh, I., Nola, E., Puca, G. A., and Cuatrecasas, P. (1973) *J. Biol. Chem.* **248**: 6543.
Sluytermann, L. A. E. and Wijdenes, J. (1970) *Biochim. Biophys. Acta* **200**: 593.
Solling, H. and Wang, P. (1973) *Biochem. Biophys. Res. Commun.* **53**: 1234.
Steers, E., Cuatrecasas, P., and Pollard, H. (1971) *J. Biol. Chem.* **246**: 196.
Steineman, A. and Stryer, L. (1973) *Biochemistry* **12**: 1499.

Tomino, A. and Paigen, K. (1970) in *The Lactose Operon* (Beckwith and Zipser, eds.) p. 233, Cold Spring Harbor, New York.

Trayer, I. P. and Hill, R. L. 1971. *J. Biol. Chem.* **246**: 6666.

Venis, M. A. (1971) *Proc. Nat. Acad. Sci. USA* **68**: 1824.

Wagner, A. F., Bugianesi, R. L., and Shen, T. Y. (1971) *Biochem. Biophys. Res. Commun.* **45**: 184.

Wankat, P. (1974) *Analyt. Chem.* **46**: 1400.

Wilchek, M. (1972) *Analyt. Biochem.* **49**: 572.

Wilchek, M. and Miron, T. (1972) *Biochim. Biophys. Acta* **278**: 1.

Wilcox, G., Clernatson, K. J., Santi, D. V., and Engelberg, E. (1971) *Proc. Nat. Acad. Sci. USA* **68**: 3145.

INDEX